# Smart Products, Smarter Services

## Strategies for Embedded Control

MARY J. CRONIN

CAMBRIDGE
UNIVERSITY PRESS

CAMBRIDGE UNIVERSITY PRESS
Cambridge, New York, Melbourne, Madrid, Cape Town, Singapore,
São Paulo, Delhi, Dubai, Tokyo, Mexico City

Cambridge University Press
The Edinburgh Building, Cambridge CB2 8RU, UK

Published in the United States of America by Cambridge University Press,
New York

www.cambridge.org
Information on this title: www.cambridge.org/9780521147507

© Mary J. Cronin 2010

First published 2010

Printed in the United Kingdom at the University Press, Cambridge

*A catalogue record for this publication is available from the British Library*

*Library of Congress Cataloguing in Publication data*
Cronin, Mary J.
  Smart products, smarter services : strategies for embedded control /
  Mary J. Cronin.
    p.  cm.
  Includes bibliographical references and index.
  ISBN 978-0-521-19519-5 (hardback) – ISBN 978-0-521-14750-7 (pbk.)
  1. Embedded computer systems.   2. Computer storage devices.
  3. Consumer goods.   4. Electronic commerce.   I. Title.
  TK7895.E42C76 2010
  004.16–dc22
  2010024609

ISBN 978-0-521-19519-5 Hardback
ISBN 978-0-521-14750-7 Paperback

# Smart Products, Smarter Services

We are surrounded by products that have minds of their own. Computing power, in the form of microcontrollers, microprocessors, sensors, and data-storage chips, has become so cheap that manufacturers are building connectivity and embedded intelligence into all types of consumer goods. These "smart products" are fundamentally changing both the competitive landscape for business and the daily lives of consumers. This book analyzes the evolution of smart products to help managers understand the impact of embedded product intelligence on corporate strategy, consumer value, and industry competition. It describes four different ecosystem strategies for designing and launching smart products: the control-focused *Hegemon*, the standards-focused *Federator*, the high-growth and brand-focused *Charismatic Leader*, and the disruptive industry *Transformer*. This ecosystem model is then applied to smart products in the automotive, wireless, energy, residential, and health industries. The book concludes with recommendations for successfully managing smart products and services.

MARY J. CRONIN is Professor of Management in the Information Systems Department at the Carroll School of Management, Boston College. Her research analyzes the intersection of business strategy and technology, with a focus on industry transformation from technology triggers such as the Internet, wireless networking, and embedded product intelligence. She is the author of numerous books, including *Doing Business on the Internet* (1994) and *Unchained Value* (1999).

*To Scott: my best chum and partner in all things*

# Contents

# Figures

# Tables

# Acknowledgments

Writing about emerging technologies and rapidly evolving industry trends in fields populated with numerous business models and relatively few success stories is a daunting task. I am deeply grateful to Paula Parish and Cambridge University Press for agreeing that the impact of embedded product intelligence on corporate strategies and individual product owners is worthy of book-length analysis. I especially appreciate Paula Parish's guidance and that of early readers and reviewers in shaping the topic to focus on business strategy and social impact rather than on the ever-shifting technical options for smart product implementation and connectivity.

As with any multi-year project, it is literally impossible to acknowledge the many experts and organizations that informed my thinking and enhanced my understanding of the technologies and industries addressed in this book. For those who so generously shared their expertise, experience, and well-informed opinions with me in the course of this project and to everyone who answered my emails, took my phone calls, endured my endless questions, recommended resources, hosted visits, and patiently explained the technical details that I needed to understand, my profound thanks.

With apologies for omitting other names that deserve to be included here, I would like to mention my appreciation for the technology, research, and business insights provided by the following individuals at different stages of my research and writing, and for the permission of some to quote from our discussions: Cynthia Artin, Daniel Bailen, Kevin Belnap, Alex Brisbourne, Patrick Byars, Tommy Childress, Ed Clark, Professor Mary Culnan, Jeanie Doty, Roger Duncan, Bob Gohn, Professor D. M. Gavrila, Ralf Hug, Professor Nicholas Imparato, Professor Sirkka L. Jarvenpaa, Joe Jumayao,

Paul Kafassis, Bob Lieberman, Letha McLaren, Robert Mazer, Paul Nagel, Michael R. Nelson, Steve Pazol, Fred Raab, Greg Rhode, Mark Robinton, David Rose, Mike Schagrin, Sascha Simon, Anthony Star, Mak Tarnoff, and Professor András Várhelyi. In addition, a warm thank you to the individuals who agreed to speak with me about their experiences and impressions as participants in smart meter and variable pricing energy pilot projects, and to the friends and family members who kindly let me explore the driver assistance, interactive options, and safety features of their recently acquired hybrid and passenger vehicles.

Thanks to the Carroll School of Management at Boston College for sabbatical leave during a critical stage of the project and for providing the support for two exceptional and talented graduate research assistants. To Peter Zeinoun, who worked with me during two years of background research, my abiding gratitude for the reports, technology specifications, copies of patents, industry data, and company briefings that you so cheerfully and competently provided. To Brian Boudreau, who arrived at Boston College just in time to endure deadlines, fact checking, citations, and manuscript revisions, I am equally grateful to you for making it a priority in your schedule to help me bring the project to completion. Any errors in this work are mine alone, and should not reflect on the help that I have received in the course of researching and writing it.

Finally, and most importantly, I acknowledge and thank my family for their support of this project and their understanding of the countless times when I had to choose research and writing over the joy of being with them. Rebecca and Johanna, thank you for listening at every stage and expressing an interest in reading the finished book even after hearing about it for years. To Scott, my endless thanks for your true and selfless partnership, for your reading and debating, critiquing and thereby improving each draft, and for the essential encouragement that you provided throughout.

# I  Evolution of embedded intelligence

We are surrounded by products that have minds of their own. Computing power, in the form of microcontrollers, microprocessors, sensors, and data storage chips, has become so cheap that manufacturers are building microcomputers and embedded software programs into all types of consumer goods. According to market research firm Databeans, microcontroller shipments worldwide will reach fourteen billion units by the end of 2010 (Databeans, 2009). Along with these chips have come a host of advanced product features and the penetration of embedded product intelligence into daily life.

Everyday appliances can now keep track of how often we use them and remind us when it is time to order new batteries or replacement parts. Alarm clocks get louder and louder, or flash a light if we ignore their morning summons. Coffee pots turn themselves on, grind the beans, and brew our first cup at just the strength we prefer. Mobile phones can download our email, display digital photos, remind us of today's appointments, and let us scan the Internet for breaking news over breakfast. Or we can start the day by listening to music on our iPod, watching the morning news reports on our high-definition television (TV), and setting up the recording of a new television series on the digital video recorder (DVR). If we forget about them in our rush to get to work, our steam irons, coffee pots, and toaster ovens will sound an alert, or simply switch themselves off to save power and avoid overheating.

Heat and air conditioning don't even need our daily attention – if we have a smart thermostat, it automatically adjusts temperature settings to match our morning departure and evening return. When we hit the road, dozens of sophisticated computer chips in our car manage the interactions among the engine, braking, steering, and other systems, record any performance problems that will need special attention at our next dealer service appointment, calculate the shortest route to our destination, and read the turn-by-turn directions aloud to keep us on track. Such encounters with embedded intelligence are barely noticeable in the first hour of a day that will be filled with similar man–machine interactions.

In fact, smarter consumer products have been proliferating for years in parallel with the increased capacity of computer chips and more recently the spread of Internet and wireless connectivity. In the past ten years low-cost microchips and ubiquitous network access have combined to bring advanced computing and communications capabilities to a long list of consumer goods. The integration of machine intelligence into our daily lives and the capabilities of today's appliances, mobile phones, automobiles, and other consumer possessions seem to have evolved quite naturally. After decades of visionary predictions about personal robot assistants, self-driving autos and home electronics that adjust themselves to our varying moods, it is hard to pinpoint any single smart product development that has burst on to the scene faster and more dramatically than anyone expected.

As long ago as the 1970s, computer enthusiasts were optimistically discussing the imminent arrival of widespread embedded intelligence and painting a generally rosy picture of its impact on consumers. One of the many books that celebrated the computer's power to transform society for the better was *The Micro Millennium* by Christopher Evans. Evans boldly predicted that by the year 2000, the integration of microcomputers into all types of products would revolutionize every aspect of our lives: "You will live in a world transformed by one tiny, cheap computer chip – the microprocessor! The microprocessor – the computer chip that processes huge amounts of information in a fraction of a second!"

Among the breakthroughs that Evans foresaw as a result of microprocessor adoption were "A pocket-size diagnostic aid for doctors, containing all relevant information; Ultra-informed machines programmed to solve world problems; Dolls and mechanical toys that respond to a spoken word; and Robots ready to cut the lawn" (Evans, 1979).

Just as today's product realities seem poised to catch up with the predictions of 1979, some planners are envisioning a future world of ultra-intelligent machines that will decisively overtake the capacity of the human brain and lead to self-directed computer power:

> Intelligent nanorobots will be deeply integrated in our bodies, our brains, and our environment, overcoming pollution and poverty, providing vastly extended longevity, full-immersion virtual reality incorporating all of the senses (like "The Matrix"), "experience beaming" (like "Being John Malkovich"), and vastly enhanced human intelligence ... We'll get to a point where technical progress will be so fast that unenhanced human intelligence will be unable to follow it. That will mark the Singularity ... I set the date for the Singularity – representing a profound and disruptive transformation in human capability – as 2045.
>
> (Kurzweil, 2005)

In contrast to living inside The Matrix and overcoming world poverty, watching TV on a smartphone or operating a robot lawnmower are bound to seem mundane. Just because today's smart products have evolved over decades, however, doesn't mean that they are less capable of upending business strategies and transforming consumer behavior. The impact of ubiquitous product intelligence is worth a closer look. It's likely that those familiar household products are already a lot smarter than you think and it's certain that they will be getting even smarter in the next five years.

Most of the fourteen billion microcontrollers shipped in 2010 are not yet connected to other devices. However, the options for adding low-cost, wireless connectivity to smart products are increasing year by year. A significant percentage of microcontroller chips already have wireless or fixed network connections and analysts predict that the number of intelligent, Internet-connected devices will reach fifteen billion by 2015 (Gantz, 2009).

Extending the power of Internet connectivity to a virtually limitless variety of embedded devices, many of them able to communicate machine-to-machine without human intervention, has far-reaching business and personal implications. Consumer products like the iPhone and the Kindle are already reshaping entire industry sectors and creating new multi-billion dollar content and service markets just as the emergence of personal computers (PCs) did in the 1980s. The hard-pressed automotive industry is counting on smarter connected vehicles to revive demand and improve profit margins. The rapid growth of global investment in smart grid and smart meter technologies is opening up power generation and distribution to new and potentially disruptive alternative energy models.

Smart products have certainly captured the attention of technology leaders and global corporations. Predictions of ever more dramatic industry and market transformation based on the next generation of intelligent, interconnected devices abound. Doug Davis, the Vice President of the Digital Enterprise Group at Intel Corporation, puts it this way:

> Generations of new intelligent devices will carry your ideas
> further into the embedded world than has ever been possible
> before, in the form of self-configuring networks of wireless
> "motes," smart factory robots, personalized in-vehicle
> infotainment systems, mobile medical diagnostic tools, and
> intelligent shopping carts. Billions of these devices will do for
> the embedded world what the Internet has accomplished for
> human beings – transcend boundaries, invent new functions,
> and redefine the way we live and work. As an industry, we
> face the amazing challenge of creating an era of embedded
> communications that will enable embedded devices to stay in
> constant contact with each other ... Extending the goodness
> of the Internet to the embedded space will have far-reaching

implications that we have only begun to imagine. Making it happen will be the ride of a lifetime.

*(Davis, 2009)*

IBM has launched a major marketing campaign to associate itself with all types of connected, embedded solutions for a smarter planet. Its white paper "Smarter Products: The Building Blocks for a Smarter Planet," presents the vision of a transformational new era of creativity and productivity based on "smarter products [that] can better adapt to the unique needs, preferences, and characteristics of consumers to get things done."

In describing how this transformation will come about, the white paper continues:

> How is this all happening? What is new? Product companies are tapping into accelerating innovation using embedded software control – the new "brains" that make products smarter. When software is effectively fused with micro-electronic, actuator, sensor, and mechanical technologies, products can become increasingly intelligent, instrumented and interconnected. That is, they can respond to changes quickly and accurately and produce better results by anticipating and optimizing for future events. They can measure and sense the relevant condition and are able to interact with other products, people, and IT systems in entirely new ways. This is what makes them "smart" – their ability to adapt to the unique needs of individual businesses and people.
>
> *(Hebner, 2009: 2)*

Much of the discussion of smart products by industry advocates assumes that new product features such as "anticipating and optimizing for future events" after the product measures and scans the relevant conditions will be welcomed and valued by consumers. Further, many companies have built business and service plans that assume smart products will command premium prices and build recurring revenue streams, creating competitive advantage. Apple's success in tightly linking the iPod device with an iTunes store, followed by a worldwide consumer embrace of the iPhone and its App store have demonstrated the financial rewards and brand loyalty advantages of controlling the content and application markets for a company's branded smart products. This success has helped to fuel a raft of competitors and imitators of the Apple smart product and content strategy. It also reinforces the perception that consumers will have a keen appetite for smart products in other market sectors.

However, the smart product value proposition for consumers is complex. In some cases industry evangelism for embedded intelligence is not grounded in an understanding of the many challenges presented by smart products and their evolving ecosystems. There are numerous examples of smart products that never captured the consumer market or that fell out of favor when lower-cost, more generic offerings came along. Consumers also have concerns about the potential negative impact of bringing more connected, intelligent products into their lives. If such products are designed "to interact with other products, people, and IT systems in entirely new ways," who will be in charge of the information that they transmit, or the decisions that they make?

Many consumers and privacy advocates are already troubled by the prospect of being surrounded by hundreds of devices that are sensing, measuring, and possibly reporting on exactly what individuals do and don't do throughout the day. What happens if the product's embedded intelligence is programmed to adhere to a built-in set of rules that its owner didn't set and doesn't want to follow? Perhaps the coffee maker is designed to use only its own capsules, meaning that it won't work with a competitor's brand of beans. Maybe the digital music that you paid to download to your iPod won't play on your new computer, or perhaps the media player embedded in your PC or your smartphone limits the number of times that certain music or video content can be played. Suppose the toaster oven you forgot to turn off sends a message to your doctor or your insurance company that this is the third instance of such absent-mindedness within a week? What if your smartphone is reporting all of your music and application downloads and Internet browsing habits to a third party that sells your preference profile to the highest bidder?

Perhaps your laser printer has automatically ordered new full-price ink cartridges even though you just stocked up at the office supply store sale last weekend. Or maybe when you stop at a garage because the "check engine" light is flashing ominously on your dashboard, the mechanic can't help you because the software to access the engine diagnostic system is only available through the auto dealership. To top it all off, you get home after a long, hot day at work and find that the smart thermostat has turned off your air conditioner in response to a signal from the utility company to conserve power. Instead of welcoming their assistance, you find yourself looking at all your intelligent possessions and wondering, "Who is in charge here?"

Not all of the projected fifteen billion Internet-connected devices produced in 2015 will be consumer products, of course. Smarter products and services are not just focused on consumer markets; robust growth for wireless sensor networks, radio frequency identification (RFID) chips, and other forms of machine-to-machine implementations will account for a large percentage of the so-called Internet of Things. As of 2010 there are approximately seven billion people in the world, but it's estimated that there are more than fifty billion machines in existence that could be connected to the Internet and to the billions of new microcontrollers shipped every year.

Before we can analyze the business and social impact of smart products, we need a more precise definition of exactly what products we have in mind and a better understanding of their technology origins and evolution. Defining the core characteristics of smart products designed for today's consumer market will be the focus of the next section.

DEFINING SMART PRODUCTS

Even in the realm of consumer products, the feature of connected embedded intelligence and the term "smart product" are used to describe a myriad of objects as disparate as automobiles, mobile phones, and electric utility meters. When applied to home appliances such as refrigerators and dishwashers a decade ago, the smart descriptor typically meant that the appliance was connected to the Internet. Now it is more likely to indicate an interface with a smart utility meter that sends the appliance signals about the cost of electric power at different times of the day and monitors its energy consumption.

The smart designation has been applied to so many products so frequently, and in so many different contexts, that it is becoming more of a marketing claim than a well-defined technical description. Which consumer devices should be categorized as smart products? While the rapid growth in embedded computing power and the absolute number of microcontroller chips are well-documented facts, the definition of a smart product is still evolving. Definitions differ considerably depending on the point of view of the writer. We will review some definitions that have particular applicability for products that target the consumer market and then provide a description of the specific product characteristics and capabilities that are most relevant for our analysis.

According to Stanford University's Smart Product Design Laboratory, the presence of an embedded microprocessor is the defining technical characteristic of a smart product: "Smart Products are products whose functionality is increased by an embedded microprocessor. It is a superset of the field that has become known as Mechatronics. Embedded microprocessors can already be found in everything from dishwashers to automobiles – and more Smart Products appear every day" (Stanford, 2009).

This is an appealingly simple definition. It encompasses the billions of microcontroller units (MCUs) that are sold into the marketplace every year and highlights specific consumer product examples. However, many MCUs are aimed at product engineers, enterprise product manufacturers, and enterprise machine-to-machine services rather than at the consumer products that are our focus. This makes the Stanford perspective too broad to pinpoint the essential characteristics of consumer smart products.

A search for narrower definitions in the field of mechatronics quickly takes us down an even more technical path. According to the editor of *Mechatronics: The Science of Intelligent Machines*, a journal of the International Federation of Automatic Control, "Mechatronics is the synergistic combination of precision mechanical engineering, electronic control and systems thinking in the design of products and manufacturing processes" (Steinbuch, 2009).

With articles like "Efficient class-B analog amplifier for a piezo-electric actuator drive," and "Application of the continuous no-reset switching iterative learning control on a novel optical scanning system," it is clear that the engineering context of the mechatronics journal is a world apart from describing smart products as they operate in the consumer marketplace.

A 2007 study on the impact that enhanced product intelligence has on consumer satisfaction provides a more accessible definition that blends embedded technology features such as sensors and software with the ability of smart products to collect and produce information of value to consumers. This article defines smart products as follows: "Intelligent products are products that contain IT in the form of, for example, microchips, software and sensors, and that are therefore able to collect, process, and produce information" (Rijsdijk *et al.*, 2007).

Researchers in the SmartProducts research project funded by the European Union provide an even more detailed definition that emphasizes the self-aware and interactive aspects of embedded product intelligence:

> "Smart Products" are real-world objects, devices or software services bundled with knowledge about themselves and their capabilities. These properties make Smart Products not only intelligible to users, but also smart to interpret user's [sic.] actions and adopt accordingly. By naming these objects Smart Products, we convey not only the notion of technology available "off-the-shelves", but also the notion of componentized software services and hardware objects required to assemble new, innovative end-user components. Therefore, Smart Products share some key properties: the ability to have multiple uses, be deployed independently, and network with others to augment their individual and collective capabilities.
>
> *(Lyardet and Aitenbichler, 2008)*

This definition encompasses the consumer product subset of those billions of objects with embedded microcontrollers that would qualify as smart products according to the Stanford Design Lab approach. It has the advantage of highlighting characteristics that make product intelligence visible and broaden its sphere of impact through networking with others; factors that have become particularly important to product designers and manufacturers in the era of widespread Internet and wireless connectivity in homes and offices.

However, this and the other definitions do not include the concept of preprogrammed control of the device, including limits on how the device can be used by its owner after purchase. As we will discuss in more depth in the following chapters, embedded controls can be fixed at the point of manufacture or can be updated and enforced through the product's network connections. The use of such controls to accomplish business objectives independent of the wishes of the product owner is an important issue. Our analysis will highlight the question of who will exercise that control once the product is purchased. Will it be the product vendor or the individual owner, or some agreed-upon combination? Understanding and resolving control issues will be a critical challenge for the future of smart products in the consumer market.

In the context of this book, smart products are defined as consumer items with embedded microprocessors and software programmed to manage various aspects of the product's functionality including any limits imposed by the vendor on how and when the product can be used by its owner. Smart products have embedded computer intelligence and preprogrammed rules for operation that govern the behavior of the product, often regardless of the wishes of its owner. Smart products are characterized by data-collection, processing, and reporting capabilities as well as connectivity to a wireless or fixed network that provides a channel for data transmission and potential for upgrades and input to the software already on the product.

In summary, this book defines smart products as network-connected consumer items with embedded microprocessors and software programmed to manage various aspects of the product's functionality.

Discussion of smart product capabilities will be a continuing theme throughout the book, starting with a closer look at the different perspectives and value propositions that the enterprise and the individual owner bring to their consideration of smart product benefits.

SMART PRODUCTS FROM AN ENTERPRISE PERSPECTIVE
This section will discuss a variety of benefits that an enterprise may consider in evaluating the potential return on its investment in developing and deploying smart products. This discussion also considers some of the risks that smart product implementation will entail. The benefits are categorized as follows: (1) outcomes that are directly positive for the company's business goals, including increased revenues, market share, and customer loyalty; and (2) outcomes that

indirectly help the company deal with social, legal, and competitive threats to its current business model.

In many cases, smart products and services are a clear extension of a company's strategy and business model. Amazon, for example, is a leading online bookseller. It has strategic partnerships with publishers, including negotiated discounts for selling their books online and agreements to offer its online customers glimpses of digitized content as part of the "Inside the Book" feature of its web pages. Amazon has also acquired BookSurge, a print-on-demand business that allows authors to self publish works and sell them on the Amazon store. Perhaps most importantly, Amazon has attracted millions of book-buying customers, creating a branded channel for reaching the book-reading market directly, in particular readers who are comfortable with online purchasing. Amazon's development of the Kindle e-book reader, the company's first foray into branding its own smart product, builds on its existing publisher relationships, customer base and online sales channel. Whether or not Amazon emerges as a long-term e-book leader, the Kindle has invigorated the long-dormant e-reader and e-book market. Amazon customers who have invested in Kindles also have an added incentive to remain loyal to the brand, since Amazon hosts back-up copies of their e-book purchases and uses a proprietary format for Kindle content that is not interoperable with other e-books on the market.

The benefits of higher-margin revenue opportunities from linking smart products and services are significant enough to attract many companies to extend into new market sectors. Apple's move to enter the digital music market was a departure from its core computer business. However, the design decisions that produced the iPod product family were clearly shaped by the same Apple product aesthetic that made its computer users so fiercely loyal. Its linking of the iPod to a branded and Apple-controlled iTunes digital music store was a bold decision to transcend the limits of the hardware-centric revenue model and establish a more digitally scalable and profitable basis for revenue growth. As of mid-2009, iTunes sales accounted for 25 percent of all music sold in the USA, with Wal-Mart, its nearest competitor, responsible for another 14 percent. And unlike Wal-Mart's sales, which include compact disc (CD) as well as digital formats, the iTunes store's purely digital operations yield higher profit margins. In digital-only music sales, Apple is dominant with 69 percent of all US sales as of July 2009 (NPD Group, 2009). These results are impressive enough to make a strong case for linking smart products and services as powerful sources of new revenue. Add in the global success of the iPhone and the iPhone App Store in attracting even more highly profitable customers to Apple and the attraction of smart products as a growth and new revenue driver becomes almost irresistible for many companies.

Amazon and Apple are two examples among many of companies that have benefited from creating smart product ecosystems. Control of the smart ecosystem allows companies to build linkages among all of the components of a smart product family such that customers will be encouraged or required to buy replacement parts or populate smart devices with content that provides a higher-margin return to product brand. Smart product ecosystem models will be discussed in more detail in Chapter 2.

Smart products may also provide bottom-line benefits by lowering the cost of after-sales product service and maintenance. This is especially important for expensive products such as automobiles and home appliances that may be covered by warranties that extend for several years. It is much less costly for the vendor to detect a performance problem and address it proactively than to send out a service representative or accept a product return. Embedded product intelligence, data collection, and connectivity features make it possible for vendors to remotely diagnose and resolve problems faster and more cost-effectively, perhaps in ways that are invisible to the product owner.

This same connected access to installed products can provide companies with valuable insights into how customers are actually using their branded smart products. The enterprise can then build on these insights to improve new product design, to supplement its product and services offerings, or to further personalize its services to create a more targeted value proposition and increase customer loyalty.

In addition to the direct benefits that smart products and services can provide to companies, they support a variety of strategies for avoiding or diminishing the impact of competitive, social, and legal threats by exercising tighter control on the use of the product and its components and content. Recent trends in business, technology, and society have highlighted the severity of external threats, including billions of dollars in lost revenue from activities such as illegal media downloads and software piracy. These trends are making threat avoidance and customer control an urgent priority for many industry sectors.

Economic trends include the recent global downturn and the longer-term tendency of decreasing prices of consumer electronics, accompanied by the commoditization and lackluster consumer adoption of higher-priced, branded consumer products. The relentless downward pressure on prices for consumer electronics has turned already low-margin hardware products into low-cost commodities. It's extremely difficult to maintain brand loyalty when consumers are intent on finding bargains, and relatively few features differentiate a high-end, branded phone, computer, or electronics device from a generic clone. Linking smart products to subscription-based services or requiring that customers buy replacement components only from the original brand provide some defense against these trends.

Laser printers that will only accept replacement ink cartridges from the original printer manufacturers are a long-standing example of using embedded controls to update a classic razor and blade strategy to the smart product era. Manufacturers can sell the printer itself at a competitively low cost and count on the recurring revenues from regular purchase of expensive branded ink cartridges to boost profitability over the lifetime of the product. Even though many of the controls designed to restrict the use of third-party substitute cartridges are possible for tech-savvy users to circumvent, the effectiveness of this approach is reflected in the sales figures for ink cartridges of original printer manufacturers compared to those of third-party cartridge sellers. The total global market for printer cartridges was estimated to be worth just over $70 billion in 2008. Of this amount, printer manufacturers sold an estimated $60 billion worth of their higher-priced branded cartridges, a significant grip on customer market share considering the strong motivation of consumers and companies to cut costs wherever possible (Scheck, 2009: B6).

The piracy of content and product cloning or counterfeit product offerings, often enabled or accelerated by the Internet and wireless technologies, is another threat for companies. These activities can significantly reduce revenue and also damage global brand value. Brands have been largely unsuccessful at controlling mass-market piracy of digital media and software, despite many attempts by entertainment companies, content owners, and publishers to create barriers to copying and distributing proprietary content. The continued escalation in software piracy and cloning motivates information technology (IT) companies and software manufacturers to seek stronger controls over the unpaid use or illicit sale of proprietary software programs and copyrighted content. The capacity of smart products and smart ecosystems to enforce Digital Rights Management (DRM) and the provisions of the Digital Millennial Copyright Act (DMCA) through embedded intelligence in the hardware as well as in the digital content offer the potential for stronger, more effective controls.

Companies can monitor and restrict customer use of content on smart devices, perhaps substituting limited term licenses and

subscription agreements for content and services previously purchased for less restricted lifetime use. Hardware-enforced DRM
using embedded controls is more difficult for consumers to circumvent; smart products can even be programmed to cease operating
if unauthorized applications or content are downloaded independently by the consumer. This also strengthens the enterprise ability to link product peripherals and services to the original smart
product, creating a tightly integrated product ecosystem. In addition to monitoring customer use of the product in order to restrict
unauthorized transfer or copying of content or new application
downloads, smart devices such as video games boxes, music players,
e-books, and smartphones may be able to remotely delete content
after purchase.

With a combination of business drivers motivating companies
to embrace smart product strategies, it is important to consider some
of the risks that such strategies entail. Many of the manufacturers of
hardware products have little experience in developing software and
linking it to services or data-collection features. Sometimes these
software capabilities are outsourced and integrated into the product
during manufacturing. In any case the new software adds complexity to the design and manufacturing process, increasing the risk of
product malfunction or failure to operate as expected. Because of
their complexity and mix of software, firmware, and hardware, smart
products have become a favorite target of hackers and criminals who
are seeking to crack these systems for their own reputation or for
financial gain. As such, smart products carry a high security risk,
and manufacturers must continually monitor and upgrade security
in response to targeted attacks.

Complex software development projects in general have a significant risk of outright failure. Even when development projects
turn out products that are ready for commercial distribution, most
newly issued software programs include numerous bugs. There is an
especially large risk associated with the increased reliance on smart
software-enabled components in complex systems such as today's

automobiles, one of the most intelligent of consumer products as measured by their use of computer chips and embedded software. In fact, many automotive recalls and malfunctions are associated with faulty electronics and software, in essence a malfunctioning of some smart component (Hebner, 2009 p. 9).

Even if a complex smart component performs flawlessly, incorporating embedded intelligence will typically add cost to designing and manufacturing the product. In automobiles, this cost of electronics and software components is now almost 30 percent of the total production cost of a passenger car and is projected to reach 40 percent by 2015 (Electro to Auto Forum, 2009). With these higher costs comes the risk that companies will not be able to recover their initial development investment because they are not able to maintain a premium price for the smarter product or convince consumers to pay subscription fees for its use over time. Many ingenious smart products designed for mass-market consumer adoption have languished in showrooms and on retail shelves. For example, only 400 consumers purchased Toro's smart lawnmower, the iMow, during its first two years in the market compared to Toro's pre-launch projections of sales in the hundreds of thousands (Black, 2002).

After a lull in the promotion of smart products after the well-publicized market failures of the early 2000s, appliance makers are again launching smart products with renewed enthusiasm, adding intelligence to large household appliances even though this particular category of consumer products has generated more than its share of market failures over the past decade. Energy conservation and government investment in a smart grid are motivating appliance brands to roll out new smart models under the banner of reducing energy consumption. In the summer of 2009 General Electric decided to take on the risk of consumer resistance to premium pricing, announcing a new model of kitchen appliances with lower power consumption as part of a major "Zero Energy" media campaign (General Electric, 2009). More cautiously, Whirlpool announced that all of its appliances would be more intelligent and smart grid connected by 2015 (Jacobson, 2009).

When smart products and embedded software combine to limit the range of choices available to the owner after purchase, for example through closed product ecosystems, there may be a negative impact on the consumer's opinion of the product, possibly reflected in online reviews and ratings. Such reactions will impede adoption of the product over time. This is especially important to consider in the context of consumer blogs, microblogging, and social networking sites where buyers share all their complaints and concerns about product performance.

This combination of smart product benefits and risks for the enterprise raises a number of questions that must be addressed in the context of the consumer view of a smart product value proposition. Do smart product features that are designed to exercise control over consumer behavior for the benefit of the vendor provide competitive advantage and increase profits, or are these controls more likely to relegate a product to a small niche market? Why do consumers welcome and reward some smart product features with rapid adoption, high sales and brand loyalty while refusing to buy and use other, arguably even more intelligent products?

The answers hinge on the company's ability to deliver a strong and easily understood value to the target customer. No matter how intelligent its products are on a technical level, companies need to position their smart products in the marketplace to highlight a compelling customer value proposition. Some features of smart products that provide benefits to the enterprise can also offer significant new value to consumers, but other aspects of smart product design are arguably not in the interests of buyers and may influence the market against product adoption. Achieving the optimal balance of value and control between the manufacturer and the consumer of intelligent products is an essential requirement for a successful and sustainable smart product strategy.

Many companies, however, seem to dedicate most of their planning and resources to internal smart product design decisions that will maximize corporate benefits and competitive advantage. Ensuring that the smart product value proposition is easy to understand and closely aligned with consumer perspectives on the benefits of various product characteristics comes later in the planning process or not at all. Consumer motivations and value perspectives are in some cases the complete opposite of the benefits that companies are seeking from smart product deployment. If most of the value of smart products is weighted towards the vendor rather than the customer, it is short-sighted to expect consumers to pay premium prices for those smart products and services when there are likely to be other options that provide more flexibility and personal control.

SMART PRODUCTS FROM A CONSUMER PERSPECTIVE

This leads us to analyze the types of benefits and characteristics that would be likely to create positive perceptions of value by an individual buyer, resulting in consumer adoption of smart products. Consumer perceptions of smart product characteristics are categorized as follows: (1) characteristics that address widely shared personal values and goals such as safety, health, and convenience; and (2) external conditions that have been correlated with consumer adoption of new technology in other studies such as network effects, ease of use, a sense of empowerment and control, and opportunities for cost savings.

Almost everyone puts a premium value on their own lives and safety as well of that of their families. This makes enhanced safety a strong value proposition for product intelligence. As with the steam iron example, where automated shut-off is triggered by sensors that monitor for overheating or lack of movement, many smart products provide safety features to protect their owners from accidents. Embedded vehicle intelligence includes smart components such as anti-locking brakes, air bags, seat belt reminders, and power steering systems. These safety options have been popular with car buyers and some are now required by national safety standards. Recent auto models feature even more advanced driver safety systems that provide lane departure warnings and collision avoidance modules. Smart product solutions for the healthcare and fitness market are also proliferating. Products featuring embedded intelligence range from sophisticated medical devices to monitor patients with chronic diseases to embedded chips in running shoes that are connected to an iPod to collect data about the runner's speed and distance covered.

Devices and services that save consumers time and make their lives more convenient are perennially attractive. One example is the worldwide popularity of global positioning system (GPS)-based personal navigation devices (PNDs) that provide driving directions and smart routing around traffic jams. However, convenience on its own may not be enough to motivate consumers to pay premium prices for smart products. PNDs became mass-market smart products when price points approached $100. At that price they were perceived as a bargain, especially when compared to the high cost of factory- or dealer-installed navigation modules in new cars. Consumer interest in improved convenience is widespread, but it is accompanied by price resistance. Evans' vision of a robot lawnmower became a reality by the year 2000, but products like Toro's iMow were thousands of dollars more expensive than standard lawnmowers, limiting adoption to a small niche of wealthy gadget lovers.

Consumers, of course, also want to have fun and stay connected to friends and family. Mobile phones, media and entertainment devices have been the strongest growth sector for smart products for the past decade. Even though these represent discretionary purchases, especially when they involve upgrading to new models, consumers have shown that they appreciate and will adopt smartphones and connected personal entertainment devices to replace older models.

Widely shared human goals, though deeply held, may not be top of mind for an individual making a smart product purchasing decision. Characteristics or market conditions that have demonstrated a positive impact on customer adoption of relevant technologies such as PCs and networked devices provide useful indicators of the consumer perspective on smart product features that do or do not represent value.

Ease of use is one such characteristic. Early Apple computers convinced consumers that it was feasible and easy for non-technical individuals to use a PC for creating personal items like birthday cards and newsletters. Smartphone touch screens and graphical user interfaces are popular because it's a lot easier to navigate and communicate with fingers and gestures than with tiny keypads. If the consumer must read a complicated instruction manual to get a new product to work, that product is more likely to end up back in the box and on the return shelf of the store where it was purchased.

Technology products can benefit both directly and indirectly from the principle that connected products and services become more valuable as they are more widely adopted, referred to as network effects. Even when relatively few consumers owned mobile phones those few wireless subscribers could benefit from the ability to place calls to landline phones wherever and whenever they wanted to. However, early adopters had few other reasons for using mobile phones. As millions and then billions of consumers became wireless subscribers, network effects kicked in. Wireless carriers upgraded network capacity and quality to handle data and media, mobile devices became cheaper and more varied, and developers started to build new and more interesting mobile applications. The value proposition of going mobile increased for all subscribers. Positive network effects from mass-market adoption also tend to increase the consumer's range of options and choice of product features and other cost–benefit trade-offs. This in turn contributes to a sense of control and empowerment in making a product selection.

Advanced technology in itself is not highly correlated with consumer perceptions of value; in combination with product complexity it can become a deterrent to adoption. Promoting products on the basis of their high-technology components and embedded microcontrollers may backfire unless the target audience is primarily tech-savvy early adopters rather than the majority of consumers.

In summary, the capabilities that make smart products particularly attractive to the product manufacturer and vendor do not overlap all that much with the consumer perspective on value and smart product adoption drivers. Unless enterprises can create a stronger business case for smart products at the intersection of corporate and consumer value, the expected enthusiastic reception of new smart product offers may never materialize.

Although the technical underpinnings of smart products are not that interesting to most consumers, managers need to understand the role of computer chips, sensors, and other embedded technology in shaping the design and core capabilities of today's smart products. The next section reviews the early applications for microcontrollers, discusses how these microcontroller foundations continue to influence the use of "hidden intelligence" in smart product design, and analyzes the impact of embedded network connectivity.

## ORIGINS OF HIDDEN INTELLIGENCE

If you ask product managers how many computer chips the average consumer is likely to encounter in the course of daily life, the answer may vary from "dozens" to "over a hundred" depending on their product domain. According to Kevin Belknap at Texas Instruments, "the average consumer touches about 300 microcontrollers every day" (Design News, 2008 www.designnews.com/article/160454–TI_Rolls_out_Low_Power_Wireless_Microcontroller_Platform.php). In a follow-up interview, Mr. Belknap elaborated on this figure with specific examples of where consumers encounter microcontrollers, starting with an estimated fifty microcontrollers in the average automobile and backed up with an explanation of the use of microcontrollers in most household appliances and all digital displays and electronic devices.

Anything that has a digital display of any kind has a microcontroller behind it, from your alarm clock to your iPod, computer and phone to thermostats, calculators, and microwaves. If there is a motor running, it probably has at least one MCU; if there is any type of sequence in terms of timing of operation (microwave, stove, dryer, snooze button on the alarm clock and so on), if it has an light-emitting diode (LED) light or screen, or if it connects with a universal serial port (USB) or has a keypad, or if one device has an impact on another like a remote control or a garage door opener, then it will be using a microcontroller (Belknap, 2009).

From a product engineer's perspective, encountering hundreds of microcontrollers every day is an obvious reality. These are the chips that have changed the face of product design. Each MCU is a self-contained miniature computing device with computational and input/output capabilities. This allows the MCU to store and execute software programs. When it is embedded in a product with other chips, the MCU can send and receive data and instructions from external memory devices and sensors, giving it an enormous flexibility and range of functionality at a very low price point. By using microcontrollers, manufacturers can create preprogrammed instructions for all sorts of devices and display screens.

The largest and most colorful exhibits at the Intel Corporation's Museum in Santa Clara, California, feature clean room bunny suits and gleaming silicon wafers. Many visitors barely pause at the detailed timeline with densely printed labels describing generation after generation of computer chip production. But Intel's breakthrough developments in the 1970s and 1980s are an excellent starting point for understanding how microprocessor and microcontroller capabilities have come to define the application of embedded intelligence to consumer product design.

According to Intel's corporate archives, the story starts in 1969 with a request from Nippon Calculating Machine Corporation, a Japanese calculator manufacturer, for the design of twelve new custom chips that would become the brains of its advanced calculator

product line. Intel engineers realized that a single chip with mul-
tiple-logic circuits would provide a cheaper and more cost-effective
design, an insight that Intel soon realized had applications far
beyond the production of calculators. Intel management bought
back the product rights from their customer and engineers got to
work on the concept of a multi-purpose chipset. Intel introduced
the 4004 microprocessor chip to the market in 1971 as "the first
general-purpose microprocessor on the market – a 'building block'
that engineers could purchase and then customize with software to
perform different functions in a wide variety of electronic devices."
The 4004 featured 2,300 transistors and could execute 60,000 operations in just one second – putting it way ahead of human calculation capabilities but providing just a fraction of the computing
power that Intel and other chip makers would offer in the coming
decade (Intel, 2009a).

By 1974, when Intel introduced its 8080 microprocessor chip,
the number of transistors had increased to 4,500 and the processing
power had soared to allow the execution of 290,000 operations per
second. This was still just a fraction of the computing power that
would fit on a chip in future years, but it was enough to confirm the
1965 prediction by Intel co-founder, Gordon Moore, that the number
of transistors on a chip would double about every two years, the fam-
ous insight that has become known as Moore's Law (Intel, 2009b).

Intel soon had competition in designing and producing microprocessors as the demand for embedded computing power expanded
from manufacturing and industrial applications to consumer devices
and telecommunications. The availability of substantial computing
power in an affordable, micro-sized package encouraged engineers
and system designers to integrate embedded intelligence into a wide
range of products and systems. As one Intel advertisement from
the 1970s described it, "Smart system designers use Intel microcomputers for almost everything, including digital controllers for
the automatic timers on traffic signals, for early video games, medical devices and home appliances" (Intel, 2009c).

Mr. Evans' 1979 vision has turned out to be much more accurate than the predictions of most futurologists. The core insight of *The Micro Millennium*, that computer chips have the potential to transform many of the products that we use in our daily lives, is undeniably true. But the impact of cheap and ubiquitous computing power has not been as universally acclaimed by consumers as he envisioned. One reason is what the display in the Intel Museum calls "Hidden Intelligence," or the development of microcontrollers.

Intel invented microcontrollers in 1976. Microcontroller design represents an advance over the original microprocessor architecture in that the microcontroller combines a core microprocessor with other elements of a typical computer system such as random access memory (RAM) and read-only memory (ROM) and an input/output capability. What's more, these embedded controllers will not typically be visible or accessible to the device owner. As Intel puts it:

> Differing from the microprocessor, microcontrollers are preprogrammed by product designers to perform the same function – whether it's controlling Ford Motor Company's automotive engines or a robotic arm from Cincinnati Milacron – and are then embedded in the application. Because of this, users often don't even know that they're there. But there they are, and in huge numbers. Indeed, microcontrollers are actually far more pervasive than general-purpose microprocessors. In 1988 microcontrollers outsold microprocessors by more than five times – with an estimated 486 million units, compared with 89 million microprocessors.
>
> *(Intel, 2009c)*

From the point of invention, then, the intended audience and primary users of MCUs were the design engineers. The primary purposes included engineering convenience and cutting product costs by embedding a low-cost chip to carry out routine instruction sets. The result was a decade of cheaper, more reliable, and more capable components for all sorts of products. This was the era when the digital display screens and timers and digital circuits made calculators and watches into cheap commodity items. But while embedded microcontrollers transformed product interfaces from an engineering design perspective, they didn't do much to make it easier for consumers to actually control the operation of their new products. Perversely, it seemed that the smarter electronics became in terms of the number of chips inside, the harder they were for consumers to operate – for example the notorious frustration of programming a home video cassette recorder (VCR) to record a TV show at a preset time, or even the tedium of resetting an embedded digital clock.

Perhaps this gap between the contributions of the MCU to engineering design and the consumer's ease of use would have been resolved in time, but another breakthrough for Intel in 1982 shifted design and development attention decisively back to the higher-end microprocessor world. This was IBM's decision to adopt the Intel 8088 as the de facto standard for IBM PCs. As a result, the 8088 was widely adopted in industrial automation, telecommunications, and military applications and products. This laid the foundation for the rise of the PC, first at the office and eventually as an essential part of the home environment. Consumers could accomplish many different things on their PCs, experiencing the impact of computer intelligence much more directly. Applications multiplied and the PC became the focal point for improvements in ease of use and end-user interface development for the next decade.

MCUs continued to evolve in the background as the hidden element of embedded intelligence. The industry as a whole continues to demonstrate the principles of Moore's Law as the capacity and performance of each generation of new chips increases while price points continue to decline. With every decade since Intel's report, microcontrollers have become much cheaper per unit and even more prevalent. The lowest-cost microcontrollers now sell wholesale for under $0.25 and even high-end 32-bit microcontrollers, the fastest-growing category, can be found for under $5.

Despite cyclical downturns in demand for chips, including the latest in 2008–2009, long-term market growth is likely to be stimulated by the increased demand for smart products and devices as well as sensor and machine-to-machine (M2M) networks. Anticipating this new growth driver, Intel and other chip makers are gearing up for the spread of embedded intelligence and Internet connectivity. Leaders of the computer chip industry are just as enthusiastic as manufacturers and consumer product brands about the potential of smart products to generate higher profit margins, even if it requires finding new ways to surface the consumer-facing potential of embedded, connected intelligence.

As microcontrollers follow the expected path of increased processing capabilities at lower price points, embedded product intelligence will become more common and even more capable in new generations of consumer products. More advanced MCUs that include a radio antenna and other connectivity features will be an important part of the projected total of fifteen billion devices connected to the Internet by 2015. It is equally certain that these smart products will be connected to other products, to wireless and fixed networks and to the Internet. These connections may be short-range linking of components to home networks and personal or environmental sensors, or they may link the product back to its vendor via the Internet, or both. The data that smart, connected products collect and aggregate will sooner or later be analyzed in order to gather insights about the behaviors of the product owner or the demographic that the owner represents. What will be the impact of adding power, connectivity, and data analytics to billions of smart products over the next five years?

Previous generations of MCUs in the home have been primarily machine-facing, used to enhance the performance or feature set of familiar appliances and consumer products with well-understood boundaries. It was a welcome evolutionary improvement when the MCUs embedded in our appliances enabled digital displays or reduced motor noise and power consumption. These were incremental improvements rather than radically different product capabilities. However, integrating network connectivity and data reporting options to MCUs that are designed for individual health, home automation, and personal energy monitoring applications represent a more dramatic shift in the role of embedded product intelligence for consumers.

Very few appliance owners want to know details about the operation of their traditional refrigerator. It keeps food at the appropriate temperature, but it has very little to say for itself and we can be absolutely confident that it's not divulging our food and energy consumption habits to any third party. With an Internet-connected, smart refrigerator we can't be so sure. If more and more of our everyday products start collecting and reporting data to third parties, that information leakage may be perfectly acceptable to the owner of such products or it may be a huge barrier to purchasing smart appliances. In either case, more consumers are going to want to know what types of information are being collected and what uses will be made of their data.

## A LOOK AHEAD

To realize the potential value of product intelligence for their company, managers need to balance a number of factors, including creation of a strategic ecosystem that will attract partners and support the core value proposition of the product and related services. Chapter 2 analyzes how companies are extending the concept of business and technology ecosystems to smart product development and distribution. A model of smart ecosystem development is presented along with examples of companies that exemplify each of the four strategies for designing and launching smart products: the control-focused Hegemon, the standards-focused Federator, the high-growth and brand-focused Charismatic Leader and the disruptive Transformer.

Embedded intelligence and connectivity provide multiple opportunities for manufacturers and brands to control the after-sale product behavior and functionality of smart products. These features also create barriers to customer exit by cutting off access to services or disabling core product functions when the buyer attempts to use the product outside of its preconfigured range of operation. As smart products begin to exercise tighter control, many brands have started to rely on capturing customers and limiting their ability to switch to other products.

Chapter 3 analyzes the most common types of smart product controls in use today and discusses how smart products that overemphasize the control function can alienate their customers and subvert market acceptance, creating limits to growth and risking an eventual backlash against the brand. The chapter ends with a model of consumer value that provides an alternative to overreliance on barriers to customer exit and customer control and supports a more open smart ecosystem model. It considers the opportunities for corporations to build a successful smart product strategy based on sharing control and generating new value for consumers. It concludes that this balanced strategy offers a path to a return on the investment that companies make in designing and deploying advanced product features, while building customer loyalty and mitigating the risk of consumer backlash and non-adoption.

Despite the recent economic downturn, automobile manufacturers remain committed to increasing the level of vehicle intelligence. As we mentioned earlier in this chapter, consumers already rely on anti-locking brakes, air bags that inflate on impact, GPS navigation systems, and many other less visible intelligent car components. Luxury cars now come with collision avoidance and smart parking options that take over from drivers at critical moments of decision. Chapter 4 provides an overview of the variety of advanced driver assistance modules that are built into today's automobiles. It then reviews strategic differences in the design and implementation of Intelligent Transportation Systems (ITS) in Japan, Europe, and the USA and discusses how these differences impact the current consumer value proposition for intelligent autos. The chapter concludes with scenarios for disruption of the automotive industry by new business models for smart hybrid and electric vehicles.

Proprietary closed systems and resistance to third-party products have a long history in telecommunications, going back to the epic battles waged by AT&T and the Bell System to block the interconnection of any third-party devices with their landlines. Until the Federal Communications Commission (FCC) Carterfone decision in 1968, customers were locked into a telecommunications market of limited product options and carrier-controlled services. A similar carrier-controls-all strategy hampered US wireless innovation in the 1990s and early 2000s. Mobile network operators set up barriers to third-party application developers and refused to support interoperable mobile devices in an attempt to control the entire mobile voice and data value chain. With nudges from government policymakers and competitive threats from new entrants, the wireless services landscape has opened up significantly in the past several years. Nevertheless, traditional wireless carriers and some new entrants are using smartphones and embedded wireless devices to control consumer behavior through newly launched smart ecosystems. Chapter 5 examines the smart device strategies of wireless carriers, mobile phone makers, and more recently media, retail, and technology leaders such as Apple, Amazon, and Google.

Like telecommunications, the electric power industry has a history of proprietary systems that interconnect only as required by regulatory authorities. Few utility companies have been open to third-party participation in energy management, much less end-user control of generating energy for themselves and for the traditional power grid. Current plans for developing a smart grid designed to accommodate alternative sources of energy and to support two-way communication with electricity customers will present challenges to utilities' mode of operation. While full smart grid deployment will take over a decade, urgent concerns about the security, scalability, and reliability of today's power grids have combined with unease about global warming to make smart energy products a top industry and government priority. In the United States, the Department of Energy has committed to a smart grid initiative that will include funding and installing two-way smart utility meters in over forty million American homes.

Chapter 6 analyzes the short- and long-term impact of smart metering on energy management options, home energy consumption, and the adoption of smart home products. It will propose that the active involvement of consumers in managing power resources and sending home-generated power back into the smart grid is necessary to create a sustainable model for shared generation of solar and wind energy and overall management of energy consumption though intelligent, interconnected products. However, the business and ecosystem changes required to enable shared control between utility companies and consumers may delay this development for many years.

For the past fifty years consumer brand companies and technology vendors have been fascinated by the concept of a smart home. The vision for this house of the future typically includes interconnected appliances that are designed to be proactive, anticipating the needs and wishes of residents. Despite decades of smart home prototypes and demonstrations, however, these futuristic home concepts are not yet part of the consumer mainstream. Nonetheless, the smart home vision is moving closer to reality as smart meters join sophisticated media and entertainment systems, PCs, wireless home networks, and smartphones in middle-class homes. Chapter 7 analyzes the benefits and downsides of smart home concepts from a consumer perspective. It reviews current consumer attitudes toward adopting home automation and notes that consumers surveyed about their priorities for smart technology in the home typically respond that they would like to simplify the operation of current entertainment and computing systems rather than add new systems. Chapter 7 concludes with a review of the leading stakeholders in smart home implementation, looking at the competition among brands and service providers to control the macro environment of an integrated smart home.

Chapter 8 analyzes why wireless carriers, machine-to-machine solution providers, and new entrants are now embracing the concept of smart services along with smart products for consumers as well as enterprise and government customers. There are critical differences, however, in positioning smart services for individual consumers versus enterprise customers. These differences are too seldom addressed in the enthusiasm for expanding smart services into the home and consumer market sectors. This chapter outlines the requirements and potential barriers for widespread consumer adoption of smart services.

As products become smarter and more connected to a network of shared data about consumer behavior, it is important that customers and vendors come to a common understanding about protection of consumer privacy. The staggering amount of data collected through wireless monitoring, combined with the ability of smart products to record and report individual activities over time, and the power of advanced data-mining tools to predict future behavior puts significant power into the hands of those who own and analyze such data. Chapter 9 reviews the types of personal data that are monitored and collected by smart product ecosystems and discusses how this data can be combined with other consumer information to reveal patterns and trends at both the aggregate and the individual level.

Current consumer protection policies and privacy regulations in the United States do not provide adequate protection for consumers interacting with smart product ecosystems. Chapter 9 concludes that smart products and smart product ecosystems have a clear potential for privacy breaches and reviews options to protect individual privacy in the smart product and services space.

From a technical perspective, it has never been easier to develop and deploy smart products. Microprocessor chips are still following

Moore's Law of increasing power and decreasing costs, making it cost-effective to add intelligence to consumer goods of all types. Embedded chips can utilize low-cost short-range wireless protocols for communications with various components of a smart product ecosystem. The product can access ubiquitous home Internet connectivity to download software updates and upload data it has collected on behalf of its manufacturer or authorized service provider.

It is far from clear, however, that adding intelligence to a product will ensure its success in the market. In fact, a successful smart product launch presents a myriad of management challenges and requires a new type of participatory product roadmap. As consumers become familiar with the interoperability limitations and higher operating costs of some smart products, they are asking more sophisticated questions about the real value of product intelligence. Customers are researching product features, reading reviews online, and checking comparison pricing sites before they buy. Social networks and blogs make it easy to share information about product performance, including concerns about restrictions, hidden data collection, and embedded controls. Brand reputations can be damaged when angry consumers use the power of Web 2.0 tools to criticize companies and spread the word about such smart product functions.

Chapter 10 addresses the issues surrounding smart products from a management and brand point of view. To help managers develop strategies for creating the next smart product success story, it summarizes the lessons learned from the successes and failures in smart product implementation to date. The chapter concludes with a model for designing and deploying smart products that encourages consumer collaboration, transparency, and open ecosystem relations rather than restrictive vendor-imposed controls. It offers strategic recommendations for successful management of smart products and smart services.

# 2 Smart product ecosystems

> The HealthVault platform supports a growing ecosystem of connected, user-friendly applications, so people can keep a comprehensive, up-to-date record of their health information in a place where they can view and share it with whomever they choose.
>
> (Microsoft, 2009a)

Combine the name of any technology company with "platform" or "ecosystem" in the search engine of your choice and you will get a sense of the maturity and scope of that company's product, platform, and ecosystem strategies. Even early-stage technology companies that don't have much of a strategy beyond issuing press releases are eager to announce an ecosystem partnership with a market leader in their industry. An August 2009 search for "Microsoft ecosystem" on the Bing search engine retrieved 6.4 million results. Most high-tech companies fall somewhere in-between Microsoft and new entrants in the scope of their ecosystem strategies. Like the concept of shaping a corporate value chain to create competitive advantage, an enterprise ecosystem has become a fundamental element in technology product development and distribution.

Applying product platform strategies to computer component design and software development gained favor in the 1990s as a way to keep up with rapid-fire technological advances and accelerating software-update cycles. Platform strategies allowed technology companies to attract more developers and component partners, to create more coherent product roadmaps, and to assure their technical and channel partners that the next generation of products would complement current development and sales plans and be compatible with their growing base of customer installations (McGrath, 1995; Salonen, 2004).

The earliest technology product platforms and ecosystems aimed to streamline innovation and facilitate interoperability. The company that owned the core technology remained at the center of the ecosystem. If that company's strategy was successful, it attracted a variety of partners who then built the products, applications, components, and peripherals that provided value-added features for the core technology. The PC partnership of Intel with Microsoft is the canonical example. The tight integration of different and complementary components that came to be known informally as the WinTel ecosystem helped Intel, Microsoft, and their ecosystem partners to shape the evolution of the PC. Intel's own product platform rolled out new versions of its chips and developer design tools at regular intervals, while Microsoft continued to upgrade and enhance its original operating system (OS) with software developer tools and suites of productivity software, allowing Intel, Microsoft, and many of their strategic partners to become the dominant forces in the computer industry and to capture the lion's share of the value generated by the ecosystem.

Microsoft ecosystem and platform partners now number in the hundreds of thousands and constitute a global economic force. According to an IDC study on the global economic impact of IT sponsored by Microsoft, in 2007 there were more than 640,000 vendors in the Microsoft ecosystem worldwide. These ecosystem participants generated more than $425 billion in annual revenues (Microsoft, 2007). In this context, Microsoft's HealthVault platform is just one of the many initiatives that have made Microsoft a force to reckon with in every industry sector.

Few companies will achieve Microsoft's level of worldwide ecosystem impact. Nonetheless, the idea that every technology company needs a range of partners and ecosystem participants to succeed in the market has taken hold. Even companies devoted to building tools for product design engineers are embracing the idea of integrating product design into a broader ecosystem. Altium, which first offered engineering design tools in 1985, reminds its customers that embedded, connected intelligence will require a new approach to electronic device ecosystems as follows:

> The next generation of electronic products won't be stand-alone
> devices, as they have been in the past. Rather, devices will be

an intelligent part of an interconnected system – a product "ecosystem" – that will offer far more than the metal, silicon and plastic from which the devices themselves are made. In essence, electronic devices are moving away from the centre of the electronics design universe and becoming satellites of a much bigger device ecosystem. The devices and the ecosystems behind them are intimately linked, and both go to make up the total user experience.

(Martin, 2009: 3)

Combining Altium's description of an integrated, interconnected smart system with Microsoft's statement about its HealthVault strategy provides a good synthesis of the key features of the smart product ecosystems discussed in this book. In our analysis, the core elements that make up a smart product ecosystem are:

- intelligent, connected devices
- closely aligned and interconnected enterprise partners who build components, applications, services, and infrastructure for these devices
- a core technology platform
- data generation and management
- a clear value proposition for ecosystem participants and smart product end-users
- participatory and user-friendly applications and services.

Although smart, connected devices have proliferated in the past decade, the other elements of smart ecosystems remain at an early stage of development. Many consumer product companies are still following an ecosystem game plan that has barely adjusted to the impact of the Internet on enterprise channel strategies. Others simply partner with the already well-established product ecosystem of a global brand like Microsoft to jump-start adoption of their own smart products or applications. But just thinking in terms of current partners and ecosystem models will limit the potential value proposition of a connected smart product for the company and for its potential customers. To take advantage of the embedded connected intelligence of smart products, smart ecosystems must create dynamic, scalable services that can extend their value proposition far beyond that offered by more traditional products and services in the market.

Smart products need an ecosystem strategy that aligns decisions about product connectivity, content, applications, product extensions, and channels for sales and support. This chapter will analyze the features of selected smart product ecosystems to provide a general framework for understanding the pros and cons of contrasting smart ecosystem models and strategies. It will discuss examples of companies that are using each model and review the benefits and risks associated with certain ecosystem decisions. The chapter concludes with a discussion of the role of the customer in smart ecosystem models as an essential and often overlooked aspect of ecosystem strategy.

For companies considering a new smart product launch, the decision about how and when to build or to join a smart ecosystem to motivate consumer adoption has far-reaching ramifications. Strong brands may decide to build and populate their own smart ecosystems, using their market leadership to establish product infrastructure and complementary services, convincing partners to adapt to the brand's technology platform and business model by providing access to a large, loyal customer base. New entrants seeking to create extended, interconnected product ecosystems to challenge market leaders may prefer to promote standards and create open-product platforms to motivate partners to develop interoperable products and services. Other ecosystem strategy dimensions that companies need to consider include the following: Does your product and platform need to disrupt one or more industry sectors and business models as a precondition to creating new revenue streams and winning customers? Which components of the smart ecosystem will be proprietary to the company and which elements would benefit most from using open standards?

If you are entering an industry sector where other companies are already well established, your market success may be premised on challenging a current market leader. On the other hand, if you are already a market leader in one or more sectors related to your smart product or service, there is little motivation to move quickly in a disruptive direction. You want a framework that builds on your existing strengths and leverages your current value chain and channel relationships. If you are a new entrant and your model requires buy-in from industry leaders, you may be especially reluctant to be seen as a disrupter of current industry business models. These types of companies may gravitate toward an incremental framework that complements their current position in the market and that provides a needed function for key strategic partners and customers.

Figure 2.1 categorizes the parameters of choice for ecosystem building on two dimensions. The first is the level of proprietary control of the participants and activities within the ecosystem, from highly controlled with participation requirements and restrictions enforced by the dominant brand, contrasted to a more open ecosystem that establishes an Application Program Interface (API) and standards as the basis for participation by a variety of ecosystem partners. New participants are welcome to join as long as they adhere to these specifications. The second dimension is whether the new ecosystem benefits more from partnering with current industry leaders for incremental smart product adoption or from disrupting the leader's revenue and business models. The next section describes each strategy in terms of this general framework for smart ecosystem analysis.

## SMART ECOSYSTEM FRAMEWORKS

Large corporations with leadership positions and an extensive base of customers for their current products will frequently gravitate towards a model for ecosystem creation that allows them to maintain strong control. Such companies often have internationally recognized brands, well-established value chains, and partner relations that support their current revenue and business strategies.

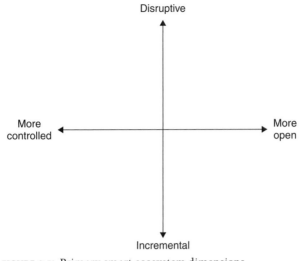

FIGURE 2.1: Primary smart ecosystem dimensions

Their market leadership position makes it challenging to introduce smart products and services with transformational potential, unless that transformation is intended to disrupt a new industry sector that the leaders aim to enter. When launching smart products within their current industry boundaries, large corporations may be inclined to build an ecosystem that preserves and enhances relationships with existing customers and partners. These are the characteristics of an ecosystem strategy that we call the **Hegemon** model.

As the name implies, the Hegemon ecosystem is controlled by one dominant company that makes the critical decisions about the ecosystem's smart products and services. This company determines the key elements of the ecosystem business model, selects the partners who will be invited to join the ecosystem and requires partners to adopt its infrastructure and platform decisions in exchange for participation. It typically uses revenues from the new ecosystem to supplement its core business model.

Since the Hegemon has an established customer base and value chain, the status quo or a gradual market evolution is seen as the best strategy to maintain existing revenues and protect market position. The Hegemon aims to develop incremental support for new smart product and service offers, since incremental smart product cycles are a good way to keep current customers and partners in sync with new product releases. Slower growth is seen as an acceptable trade-off for avoiding marketplace disruption. To retain control of the technology, the Hegemon must be willing to invest heavily in research and development (R&D) because it will need to set the pace for innovation within the ecosystem.

The strategy of evolving a smart ecosystem around a company's existing products and value chain relationships has obvious appeal for established market leaders. It keeps the company in control, positioning their own products, services, and technology roadmap at the center of the new ecosystem. The lead brand can stage the rollout of interoperable services to avoid disruption or cannibalization of its current business model. It is also very tempting for large companies to try to capture the lion's share of any new revenues that may be generated by future growth of the smart ecosystem and to exclude potential competitors from any meaningful participation.

If, on the other hand, a well-established company is entering a new industry sector where its primary challenge is to gain market share quickly, then that company is likely to consider the advantages of a different ecosystem strategy that we call the **Charismatic Leader** model. When entering a different industry sector, established brands may be more willing to embrace a disruptive approach that will directly challenge existing market leaders and traditional industry business models. The entering brand needs to convince its current customers to adopt the brand's smart products in a new and often highly competitive market. Strong customer adoption will demonstrate to early ecosystem partners and prospective participants that the Charismatic Leader has the clout to prevail against entrenched industry offers.

As the name implies, the Charismatic Leader ecosystem model is centered on the power of a leading brand to win a strong, enthusiastic following of customers and ecosystem partners for its smart products and services. Both the Hegemon and Charismatic Leader models provide avenues for strong brands to maintain control of core components of an emerging smart ecosystem. The Hegemon is betting on its ability to evolve the system over a long period of time while it sustains its current business model and proprietary value chain partnerships with a few hand-picked partners. The Charismatic Leader, on the other hand, is betting that the strength of its brand and business model in one industry sector will enable it to enter another sector with a different smart product and achieve similar levels of customer adoption in a relatively short period of time. The Charismatic Leader aims to exercise control over its own ecosystem partners and service providers while actively disrupting the current business models and market leaders in the new industry sector.

To succeed in disrupting market leaders and business models, the Charismatic Leader will need to have a large, enthusiastic, and highly loyal customer base. But the ability to pull its current loyal customers into the target market sector isn't enough. It will also need to take market share and customers from the sector's current leaders. This requires creating a qualitatively different and better experience and a wider set of options for customers. Ideally the ecosystem will succeed in enhancing brand value by attracting new customers through smart product and service innovation.

Compared to the Hegemon model which is selective about approving ecosystem participation, the Charismatic Leader needs to recruit large numbers of partners who will provide interoperable services and content for its smart product platform as quickly as possible. Leveraging the innovative capabilities of the entire ecosystem community will help enable the brand to stay ahead of its more traditional competitors. However, the Charismatic Leader is not willing to give up control of certain key aspects of the ecosystem. Too much openness may damage its current brand image. The brand needs to convince prospective partners and customers that they are indeed benefiting from being early adopters of this branded smart product to justify the level of control that it intends to exercise within the ecosystem.

A third smart ecosystem option, called the **Federator** model, combines incremental growth strategies with a higher degree of openness for participants. In the Federator model, the ecosystem is built around a core technology or specification that the creator is promoting as a platform for a variety of smart products and services that can be developed by partner companies of all sizes. For market-leading partners, the Federator platform adds a new capability to existing products and is typically designed to fill a need for interoperability among multiple competing products in the industry sector. For new entrants, the platform will give them a jumpstart on entering the market and will assure customers that they are providing products that are interoperable and can be relied on. The central technology platform in a Federator model is often based on specifications designed by the ecosystem founder, then donated to a consortium, standards body, or industry alliance as an open platform. Federators may also be technology platform providers looking to license their solution as an advanced and stable solution in still-emerging industry sectors. In that case, the platform itself may remain proprietary as with Microsoft's HealthVault for patient health and medical records. Federators may also adopt specifications or standards that are independently developed by established industry consortia or through formal standards bodies. Companies that are not yet well-recognized brands and do not have a loyal customer base or consolidated leadership in their current market space may rely on such standards to attract both well-established market leaders and innovative smaller companies to join with them.

Recruiting new ecosystem partners to a Federator model is often a slow process. It may take years for a critical number of partners to integrate their products with the central platform or specification. In theory, the Federator model is designed explicitly not to favor one brand over any other participant, or to allow a single brand to determine how other participants build and market their own products. Its value proposition for partners is interoperability and stability. Rather than embracing a disruptive business or revenue model, the Federator creating the ecosystem is expecting that many participants in the new ecosystem (especially those partners who have already built up a large product family and customer base) will incorporate the ecosystem platform into their existing business and value chain models. That takes time and resources. For these reasons, Federator founders must make a long-term commitment to incremental growth. Many large corporations adopt the Federator model as a path to extend their technology platform into new industries once they are well established in their original market.

Every new partner that integrates products with the platform will reinforce the momentum of market adoption and make the ecosystem more viable. So it is in the interest of the Federator to reassure ecosystem participants that the platform will support and extend their current business models. In this sense, the Federator model is allied with the industry sector power structure. A Federator is eager to recruit companies to declare support for the standard and build smart products and services using its platform and interoperability specifications. Anyone is welcome to join the ecosystem as long as they are compliant with the adopted specifications and interoperable with other products. This interoperability is a strong selling point of the Federator model because it creates network effects once large numbers of products have been developed and sold to consumers.

The **Transformer** is the fourth and final ecosystem model in this framework. Transformer ecosystem founders tend to be more open than the Hegemon and Charismatic Leader companies. They are also quite willing to disrupt the status quo and to challenge current market leaders in order to gain market share for new entrants. Instead of courting strong industry partners like the Federator, the Transformer is willing to work outside of current business models in order to develop new ways of delivering services and monetizing smart product adoption. Transformers may be large corporate entities with strong brands who are interested in using the ecosystem to energize their primary business models, or they may be new entrants with an advanced technology, an aggressive pricing strategy, or innovative distribution policy designed to challenge established industry product and service categories. Often the founder of a Transformer ecosystem has already established itself as a leader in another industry sector. Its core business model generates revenues that are independent of the established channels for sales in the industry sector that it aims to transform.

Figure 2.2 illustrates the four ecosystem models with an example company within each quadrant.

Which of these models is the best approach for a company planning to create its own smart ecosystem to launch a new smart product and related smart services? Each model has its own advantages and risks. The choice depends on that company's willingness to open its product platform as a standard API to attract ecosystem partners, together with its appetite for disrupting the current market leaders. Before drawing conclusions about the strengths and weaknesses of the four models, the next section describes each of these ecosystem strategies in action, starting with the General Motors (GM) OnStar solution as an example of the Hegemon model.

### ECOSYSTEM MODELS IN ACTION

*Hegemon model*

The Hegemon model is an appealing choice for companies offering products and services where safety, performance, stability, and security are at a premium in the mind of the consumer. This clearly applies to automobiles; a very high-stakes purchase because of the high initial price along with the ongoing requirements for service and maintenance throughout the life of the vehicle. The warranty depends on the continued existence of the manufacturer who must stand by any required repairs that are a result of design and component flaws. The need for consumers to rely on the business stability

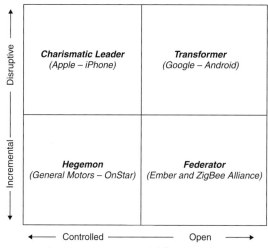

FIGURE 2.2: Smart ecosystem model framework

and reputation of the manufacturer for the life of the vehicle as well as the intrinsic safety concerns about auto performance, combined with the complex and expensive value chain structure required for auto makers to produce vehicles, make it likely that traditional auto makers will favor the Hegemon model.

The GM OnStar solution provides a good example of the rationale and the pitfalls of creating a new smart product ecosystem that is based on a Hegemon control strategy.

GM collaborated with EDS and Hughes Electronics to develop the first OnStar system in 1995. It leveraged existing auto technology components to create a number of innovative services for its customers. The OnStar emergency assistance service was based on sensors inside the vehicle's air bags and an embedded GPS location module. GM linked these sensors to a telematics network that would notify a central OnStar office when deployment of the air bags indicated a serious automobile accident. The OnStar service would then alert local emergency services of the vehicle's location based on the embedded GPS. In another service, offering an engine performance monitor, GM surfaced the data from the embedded On Board

Diagnostic (OBD) system to inform drivers about potentially serious engine problems. At the time it was launched for selected models of the 1997 Cadillac, OnStar represented significant advances in the location and communication services available for passenger cars. OnStar subscribers could also use the service to proactively request assistance during a less serious roadside emergency such as a minor accident or a flat tire.

In marketing the OnStar service, GM opted for an incremental path that would support its current value chain as well as enhance its brand reputation and provide a new source of recurring subscription revenues. Installation and service activation were initially carried out at the dealership, rather than taking place at the factory. GM's target sales for the launch year were modest – activation of OnStar in 100,000 vehicles. But OnStar installations did not achieve even this conservative sales goal. Dealerships were not effective sales channels for the innovative value of the advanced but unfamiliar features that OnStar provided, especially at the price point set by GM for the installation and activation. After a few years of slow growth, GM celebrated the milestone of 50,000 subscribers in March of 1999 (GM Media OnLine, 2009).

At this point, GM and OnStar executives rethought their production and sales strategies to accelerate customer adoption. Rather than shifting away from a Hegemonic model, they decided to implement factory installation of the embedded OnStar components during manufacture, with car buyers encouraged to activate the service features at the point of sale. Factory-embedded OnStar systems were less expensive for GM to install, enabling the company to incorporate OnStar into more of their vehicle models. To familiarize the buyers of vehicles that now came pre-equipped with OnStar options with the full value of the service, GM began offering new car buyers a year of free OnStar service before requiring a subscription fee. This strategy resulted in rapid expansion of OnStar-enabled vehicles and in April 2001, OnStar celebrated the milestone of one million subscribers (cited from GM Media Online, 2009).

GM also worked to expand the range of services provided by OnStar. New versions of the service incorporated cellular connectivity, with an embedded dashboard control panel that included a Talk button that enabled wireless voice-activated calling and a speaker-phone component. Instead of connecting to the driver's own mobile phone, the calls were handled by a wireless carrier ecosystem partner that provided wholesale calling minutes for resale to OnStar subscribers. A 2001 partnership with ABC News, ESPN.com, and Disney.com delivered in-car news and entertainment services. Voice-activated, turn-by-turn navigation assistance was provided by a centralized call center staffed 24/7 with advisors who responded to driver requests for directions. These service decisions cemented GM's business model as a relatively high-priced subscription option, with monthly subscription fees of $18.95 per month for the emergency assistance "Safe and Sound" plan and $28.90 per month for "Directions and Connections" which added the turn-by-turn navigation and information services to the Safe and Sound option. To use the voice-activated calling plan, subscribers had to pay separately for a calling-plan package and the voice minutes they used.

Even though GM continued to add services to OnStar over time, its internal technology development struggled to keep pace with the major advances in wireless and GPS technology in other industry sectors. A decade after its launch, many OnStar services were replicated by stand-alone products that provided similar and often significantly lower-cost solutions. Cellular coverage and reliability improved nationwide, the majority of car owners became mobile subscribers, and the personal mobile phone was enough of an emergency assistance tool for most drivers. Mobile handsets became more sophisticated with built-in speaker phones or low-cost battery/charger speaker-phone attachments that reduced the appeal of subscribing to additional minutes through OnStar. Personal navigation device (PND) prices fell to below $200 and the GPS features on these stand-alone devices expanded to include full color maps with additional information about local attractions and even peer-to-peer map updating and commentary as well as spoken turn-by-turn directions. The OnStar voice-only navigation guidance went from being advanced to seeming a bit old-fashioned. It was hard to convince car owners that they needed to invest hundreds of dollars each year in an OnStar subscription service when many of its features were already at hand in other smart products.

The value of GM's investment in the OnStar ecosystem and its portfolio of services was in danger of being eroded, reducing their technology advantage to those components that could not be easily replicated by a non-embedded, stand-alone device such as PND or mobile phone. One such feature debuted in 2008 as the vehicle recovery service. With the car owner's authorization, the police can activate an engine control option in OnStar that blocks the car from starting, or if the car thieves are already on the road slows down the car gradually until it is only running in idle mode, allowing police to apprehend car thieves while avoiding potentially damaging high-speed chase scenes.

The OnStar website emphasizes the high satisfaction level of regular OnStar users and the especially positive response to the inclusion of the engine-block disabling feature as part of the basic service package (OnStar, 2009). But using OnStar to block vehicle ignition and degrade engine performance from afar has highlighted long-standing concerns about the control that car owners may be giving up when they subscribe to the service, along with renewed questions about the privacy and ownership of the data that OnStar routinely collects. Year to year OnStar subscriber growth suggests that the number of GM car owners who opt to pay for the OnStar service after the first year of free availability remains low. In the last quarter of 2009, OnStar had 5.5 million subscribers, including those still covered by the first year of free access after purchase. This was far less than half of the potential subscriber base of over sixteen million OnStar-enabled GM vehicles sold (Migliore, 2009).

After emerging from bankruptcy in 2009, executives at GM asserted a continued commitment to OnStar, but at this point its once pioneering smart service is competing in a much more technically advanced, connected, and crowded smart product market. The OnStar services and development cost structure is high relative to the top-line revenue that it generates. As long as it remains committed to the Hegemon model, GM will find it challenging to find new avenues for revenue growth.

*Hegemon model risks*
In exchange for keeping such tight control of the ecosystem technology, the Hegemon must bear the bulk of the required investment in R&D to add new features over multiple generations of its platform as the market demands and competitive offers become more sophisticated. This approach puts the Hegemon at risk of falling behind while competitors leverage more open-technology options and ecosystem partner innovations to create smart products and services on top of a more dynamic shared platform. Since development and innovation cycles have shortened dramatically with the availability of open systems and ubiquitous connectivity, the slower pace of Hegemon ecosystem development becomes a bigger risk over time as many aspects of the Hegemon's previously unique offer are more easily replicated. In GM's case, the OnStar combination of telematics, cellular networks, and GPS satellite location devices with turn-by-turn directions and mapping make it especially challenging to keep abreast of advances in each of these systems – for example the advance from 2G to 3G to 4G cellular networks and the proliferation of smart mobile handsets with GPS receivers in the USA and Canada.

In addition, the reluctance to disrupt current value chain relationships and business models creates the risks described by Christensen in *The Innovator's Dilemma* – being blindsided by disruption from new vendors who enter from below or from outside of the industry sector (Christensen, 1997).

Price points of new entrants for smart, connected in-car navigation and entertainment devices are significantly lower than OnStar's established pricing structure, illustrating the risk of trying to extract premium prices from customers in the Hegemon model. Consumers are free to select after-market vehicle enhancements based on considerations such as value-added features, pricing, and openness of the platform to innovation, personalization, and peer participation. GPS device makers have capitalized on this factor to differentiate their product offerings and to develop ecosystem strategies that position them to be industry Transformers by disrupting the market for higher-priced factory-installed navigation assistance.

One way to address these risks is for a company that starts with a Hegemon ecosystem to expand their tightly controlled partnering model and shift to a different ecosystem model over time. The Federator model offers opportunities to attract more partners and generate revenues from licensing the core technology. OnStar executives experimented briefly with the Federator approach to customer growth by licensing the OnStar technology platform to other auto makers starting in 2002, but it ended those licensing agreements after a few years. The Hegemon strategy of keeping tight control of the ecosystem hindered management from adapting the service to compete more effectively. In the face of economic pressures in 2009 GM executives indicated that licensing OnStar was back on the table for discussion (LaReau, 2009). After many years of branded platform control and dominance, however, it will be challenging to convince competitors that licensing the solution will be advantageous. And even if the core products are still technically advanced enough to be attractive for licensing, it will be difficult for GM and OnStar to give up the degree of control it has been exercising over its ecosystem.

Moving towards a Charismatic Leader model is even more challenging because it requires a willingness to transform industry business models and value chain relationships, which may create negative impacts for current ecosystem partners as well as for the Hegemon's own revenues. In addition, dealing with a much larger number of ecosystem partners and leveraging their participation to significantly broaden the range of products and service offerings while still exercising control requires a complex skill set that typically has not been developed in managing the Hegemonic ecosystem.

## Transformer model

The widespread availability of stand-alone GPS location devices and low-price PNDs has transformed how millions of drivers around the world find their destinations. Over 41 million PNDs were sold worldwide in 2008. As is often the case with a new and disruptive technology, the PNDs have limitations compared to embedded systems such as OnStar – they cannot provide the automatic, driver-independent emergency service calls based on accident detection that GM offers to its subscribers and they don't provide access to a live person who can answer questions as well as giving directions. However, with the rapidly decreasing cost of PNDs, consumers seem inclined to adopt them as a low-cost and generally adequate solution for their navigational needs. Aggressive discounting in the USA brought retail prices down to $100 for basic models by the end of 2008, making this device a very affordable option for all drivers.

That price point is a fraction of the cost of the higher-end embedded navigation systems which average over $1,000 for new cars. In the context of this dramatic price difference, it's easy to understand why the number of new car buyers paying for an embedded auto maker-installed navigation system remains below 20 percent even though such systems are now an option on over 80 percent of new passenger vehicles (Newcomb, 2009). In terms of the automobile navigation market, PND makers have succeeded in disrupting the business model of OnStar subscription services and the profit margins for auto makers investing in embedded GPS systems.

As vendors entering a new market space, PND manufacturers adopted the Transformer model as the fastest way to win market share and create revenues. Each vendor established its own ecosystem partnerships for development and distribution. Garmin, the PND market leader in terms of devices sold per year, started out in 1990 to manufacture and market specialized GPS devices for the hiking, aviation, and boating markets. It began making automotive navigation devices in 1997 and by the time the car owner demand for PNDs accelerated in 2003 it was able to use its worldwide dealer and distribution channels to quickly gain market share. TomTom, second to Garmin in market share, has been focused on car navigation and digital mapping since its start in 1991.

As the cycle of technology and smart product innovation keeps moving faster, the companies that manage to transform one sector's business model can find themselves threatened with disruption by an even newer smart product and service platform. That is happening in the personal navigation space. Makers of dedicated PNDs such as Garmin and TomTom find themselves being squeezed by the recent popularity of smartphones with large screens, GPS chips, and downloadable navigation applications. Smartphone applications can provide easily readable maps and turn-by-turn directions for drivers and have the added advantage of being always connected, making them readily updated to reflect changes in road conditions and even real-time traffic bulletins, and able to summon emergency roadside assistance in most instances. In October 2009, Google announced the availability of a free navigation application for the Android smartphone platform that provides many of the features of stand-alone PNDs.

One advantage of the Transformer model is that the ecosystem lends itself to updating product platforms, expanding services, and even changing business models much more readily than the Hegemon model. PND makers are reconfiguring and enhancing their devices to compete more effectively with smartphone-based services, adding embedded connectivity to their PNDs, and value-added features such as smart routing advice based on the shared experiences of millions of existing customers, news, information and entertainment content partnerships, and location-based social networking.

TomTom, for example, is enabling real-time connected services by putting a cellular Subscriber Identity Module (SIM) chip in its premium PND product line. TomTom is also encouraging customers to take a more active role in shaping navigational features by asking them to upload logs of their past trips to create a new smart service that can take traffic patterns at different times of day into account when providing directions. ABI Research estimates that 34 percent of PNDs will feature real-time two-way connectivity by 2013, paving the way for a further expansion of connected services on the PND platform (TomTom, 2009).

To counter the threat from GPS-enabled smartphones, TomTom has developed its own mobile navigation applications and is selling these from mobile application stores. Like other PND makers, it is also partnering with auto makers that are now interested in opening their ecosystems to partner platforms to provide lower-priced embedded GPS options on selected vehicles. TomTom and Renault have collaborated on the Carminat™ system in Europe, an indication that Transformers are in a strong position to extend their ecosystem strategies into other models (TomTom, 2009). Garmin is also working with car makers on embedding connected navigation devices into their vehicles. It has also moved into the mobile handset space with the release of a Garmin-branded phone in partnership with ASUSTeK (Garmin, 2009).

The Transformer model is well suited to companies prepared to keep pushing ahead with technology innovation, whether they are new entrants or global companies that have established their brand and business model in another industry sector. Well-established Transformers can try out disruptive business models in a new market space, opening up that space as a new source of revenues if their model succeeds and gaining useful strategic experience even if it falters. This option makes the entrance of Transformers a constant threat to market leaders in industry sectors such as wireless, energy, and healthcare where smart products and services are rolling out at a fast pace.

Google's introduction of the Android™ operating system (OS) for mobile devices is one example of a well-established Transformer taking aim at leaders in a different industry sector. Google didn't have any prior track record in mobile devices or in operating systems for smart products and as of 2009 Google as a company still does not manufacture or brand any smart products of its own. Nonetheless, it has succeeded in establishing a smart ecosystem for Android based on the Transformer model. The Android OS for smartphones is available to phone manufacturers and to makers of other smart devices such as netbooks free of royalty and licensing charges, making it a cost-effective option for manufacturers. To make the Android ecosystem attractive to application developers, Google released an early version of the Android smartphone software development kit (SDK) and sponsored several contests that awarded monetary prizes for innovative applications. It invested in creating an Android application marketplace where developers can give away or sell their downloadable Android applications without prior permission from Google or a wireless carrier, a radical departure from the traditional walled garden model of carriers and the highly controlled app stores created by Apple for the iPhone and RIM for the BlackBerry™. Although the application developer community around Android hasn't grown as quickly as that of Apple's iPhone, it is already sizable and poised to take off when more wireless carriers and manufacturers bring out Android mobile devices.

The first handset based on Android, the G1, was manufactured by HTC and launched by T-Mobile USA in October 2008. Behind the scenes, Google's work with ecosystem partners was demonstrated by the growing ranks of members of Open Handset Alliance (OHA), the association that officially supports development of the Android operating environment. OHA members in 2009 included the world's largest wireless carriers such as China Mobile, Vodafone, NTT DoCoMo, Telefónica, T-Mobile, and Sprint, along with device makers including HTC, Garmin, ASUSTeK Computer, LG, Motorola, Samsung, Sony Ericsson, and Acer. This roster of Android ecosystem partners is rounded out by a number of the largest semiconductor firms and a variety of firmware and software companies.

Apple, Nokia, Palm, RIM, and Microsoft are all notably absent from the OHA. Android is designed to disrupt their existing ecosystems' operating system strategies. Historically, OS licensing and device configuration restrictions have been used to control mobile device features by keeping most firmware designers and application developers away from core OS functions and network-facing capabilities. Google's decision to open this access to third parties using Android could significantly accelerate embedded mobile development, providing more innovative devices for consumers, speeding the pace of firmware implementation and the potential for convergence of portable, embedded mobile devices with radically new applications and features.

This potential for transformation increases as more participants join the Android ecosystem and begin to develop their ideas and inventions; it is this open Android model that Google is counting on to create a sustainable advantage. In an interview with CNet, Andy Rubin, the director of mobile platforms for Google, says, "A single product is going to have, eventually, limitations. Even if that was two products that's going to have limitations. But if it's a hundred products, now we're getting somewhere, to the scale at which Google thinks people want to access information."

Reflecting on the advantages of open source and an open ecosystem approach for Google, he adds, "Getting back to business models, Google has a great business model around advertising, and there's a natural connection between open source and the advertising business model. Open source is basically a distribution strategy, it's completely eliminating the barrier to entry for adoption" (Krazit, 2009).

By the end of 2009, the results of Google's strategy were beginning to pay off. With announcements that Motorola would use Android as the OS for most of its new handsets and several other manufacturers committed to rolling out new devices based on Android in 2010, it seemed that Rubin's goal of hundreds of Android devices was turning into a reality. Reflecting on the advantages of an open standards approach and a rapidly growing group of ecosystem partners, Juniper Research predicts that by 2012, the number of Android devices in consumer hands will far outpace the number of iPhones sold. Juniper notes that open source operating systems are already playing an important role in differentiating smartphones and predicts that open OS phone shipments will increase from 106 million units in 2009 to 223 million by 2014 (Bethell, 2009).

The Android platform is also positioned to become the device platform for netbooks and other personal computing devices, making Google the potential center of an even broader ecosystem where every Android device can be optimized to access related Google services from search to maps to Google applications in the cloud.

### Transformer model risks

Transformers must be able to thrive in an environment of industry disruption. An important success factor is the ability to constantly innovate and if necessary cannibalize one's own business models. If a Transformer loses that innovative edge, it will be vulnerable to other aspiring disrupters. Disruptive models based on open platforms can grow quickly once they are established, but before new standards, alliances, and products can be built, there will be a considerable time lag. It may be years between the launch of an innovative technology platform, winning new partners to join the smart ecosystem, and the point when customers adopt the resulting smart products and services. That may require the Transformer to fund incentives and support for ecosystem participants to join and become active before the customer base is big enough to generate meaningful revenues.

The more open the Transformer ecosystem is, the greater the risk that some participants will not adhere to quality standards, degrading the overall value of the ecosystem. Transformers typically lack the enforcement mechanisms for quality control provided by the founders of Hegemon and Charismatic Leader ecosystems. Unlike Federator ecosystems, Transformers do not necessarily have a formal process to certify that products interoperate with each other.

The quality of new smart products and partner solutions is therefore likely to be uneven. If too many developers and device makers take short cuts, it may alienate other partners as well as consumers, and interoperability and reliability problems will develop.

Transformer ecosystems must ensure the rapid growth and customer adoption needed to keep all participants engaged and appropriately rewarded for their work on the new platform; if they don't attract enough end-user customers to generate revenue for the ecosystem participants, they will be relegated to a niche position in the market rather than truly disrupting the current industry leaders.

*Charismatic Leader model*

Is it possible for a brand to maintain control of key elements in its product development and distribution strategy and also establish an ecosystem that fosters – and benefits from – rapid innovation among thousands of partners and new ecosystem entrants? Five years ago, that might have seemed impossible. But the emergence of smart, connected products has enabled new ecosystem models that combine brand control and disruptive innovation.

As a new entrant into mobile devices and wireless services, Apple saw the need to break through traditional wireless industry strategies. The iPhone had to disrupt the Hegemon models of wireless carriers who maintained a lock on mobile device approval as well as a walled garden environment for applications and subscriber services. With an eager group of loyal iPod and Macintosh customers waiting for the release of the first iPhone handset, and a highly profitable iTunes distribution platform, Apple pushed for carrier business agreements that broke with previous handset deals. Apple maintained control of the iPhone ecosystem to the benefit of subscribers and developers, creating high-profit revenue opportunities, including the sale and activation of iPhones directly by Apple in its retail and online stores, approval of all applications submitted to the Apple-branded App Store, and control of the distribution and sale of iPhone applications directly to subscribers.

The deal with carriers was only one step, however. Apple also had to convince application developers to sign on to the stringent control requirements of the App Store, which include Apple's review and approval, or rejection, of every application that is submitted. Apple won developer participation by bringing a critical mass of enthusiastic customers to the App Store, and offering developers a revenue share agreement that made it profitable for small and large developer groups to focus on iPhone apps – even at the price of ceding control of key decisions about app approval to Apple.

It's easy to declare yourself a Charismatic Leader in an ecosystem model, but that alone is not going to ensure that you will find customers who are eager to adopt your new smart product offering because of their devotion to your overall brand. Microsoft found that out with its launch of the Zune mobile music player in 2006. In the three years since its introduction, Microsoft has managed to sell 3.75 million Zune players, representing about 1 percent of the music player market compared to Apple's 74 percent share with its ubiquitous iPods (Kramer, 2009).

Nokia, a powerful global force in the wireless industry, has repeatedly attempted to launch the type of iconic smartphone design that catapulted Apple into mobile device leadership. Nokia sold 468 million mobile handsets in 2008, but its market share has slipped in the past year, particularly in the highly profitable smartphone category. Nokia phones are often the low-budget choice for consumers who tend to ignore the company's release of more ambitious high-end gaming, entertainment, and productivity devices. It may be that Nokia's market leadership and dependence on long-standing carrier partnerships are part of its problem in transitioning to a more disruptive Charismatic model.

Nokia is looking for ways to enter new markets and seems willing to play a more disruptive role in other industry sectors. In its recent entertainment initiative "Comes with Music" Nokia has attempted to outdo Apple in its music download sweet spot by integrating an unlimited, prepaid license to listen to digital music for buyers of selected Nokia handsets. After some hesitation, major record labels have signed on to this opportunity in return for collecting some guaranteed advance revenue in exchange for giving up per track download fees. To attract more customers and ecosystem partners for content and application development, Nokia has also revamped its content distribution ecosystem and launched Ovi, a Nokia-branded mobile app store. At the same time that Nokia is going after a slice of new business in the mobile applications and music sectors, it is trying on the role of Federator for mobile device operating systems, opening its previously controlled Symbian operating system. Nokia launched Symbian in 1998 as a joint venture with Ericsson and Psion to create a mobile operating system that would be used in new Nokia handsets, and licensed it to a limited number of other handset manufacturers who paid a six-figure membership fee to join the Symbian Alliance. As of June 2008, Symbian became an open software platform available under a royalty-free license to members of the Symbian Foundation willing to pay a much lower fee of $1,500 annually (Symbian Foundation, 2009).

Nokia's efforts to expand into new sectors have created a somewhat awkward hybrid strategy without generating the groundswell of consumer enthusiasm for new smart product releases that is characteristic of a successful Charismatic ecosystem model. There are too many competitors who can turn out comparable low- and mid-range mobile handsets, and more competitors waiting in the wings who are willing to make the concessions that wireless carriers typically demand. As a market leader in the wireless industry, Nokia may not be willing to embrace significant disruption that would endanger its leadership position with carriers and existing ecosystem partners. Nokia is dependent on wireless carrier business for its primary revenue. So Nokia is caught between a somewhat bland product image with consumers and a long-standing relationship of collaboration with wireless carriers, both of which make it difficult to obtain disruptive concessions for revenue sharing and device configuration control. Despite its current problems in losing market share to Apple in the profitable smartphone sector, Nokia remains a force to reckon with. In the first half of 2009, it still commanded an impressive 38 percent of total mobile handset market share worldwide (Nokia, 2009).

Amazon's strategy for the Kindle, its branded e-book reader, closely matches the ecosystem model of the Charismatic Leader. Even though Amazon has remained inside the publishing and bookseller industry sector with Kindle, it seems well positioned to disrupt the already shaky print publishers' business model and to create a wirelessly connected e-book and downloadable content ecosystem. Thanks to its online store, Amazon is the dominant online bookseller by a huge margin. As of 2008, Amazon became the largest retail bookseller in North America as well as internationally, with North American book and media sales totaling $5.35 billion; overtaking the second place retailer, Barnes & Noble/B. Dalton, which sold $4.52 billion in books and media during 2008 (Rosenthal, 2009).

So when Amazon decided to shake things up by demanding that publishers sell their e-book versions for a discounted price and in a specialized format for the Kindle, it didn't risk destroying its current relationship with the industry leaders, at least in the short term. Publishers need Amazon.com at least as much as Amazon needs them. Some have voiced their resentment and a few have resisted the aggressive discounting model, but for the most part publishers have agreed to join the Kindle ecosystem on Amazon's terms (Hall, 2009). Publishers may be eager to make deals with other e-book platforms that support more profitable pricing models, but as long as Kindle is successful in attracting book buyers and publishers are dependent on Amazon for print book sales, these partners will stay in the fold.

Like Apple, Amazon wields the power of a large and loyal base of customers for Amazon.com. Amazon does its best to shroud the data about Kindle sales in secrecy, giving general indications about device sales figures and focusing on the number of e-books that have been sold. One analyst estimates that about 1.5 million Kindles had been sold by the end of 2009 and predicts that about 1.8 million more units will be sold in 2010 (Savitz, 2009). Analyst extrapolations from available data about Amazon's e-book sales indicate that Kindle buyers are avid consumers of digital content and that by 2012, Kindle will generate $2 billion a year in revenues for Amazon (Barron's, 2009, http://blogs.barrons.com/techtraderdaily/2009/06/04/amazon-2byr-kindle-sales-by-2012-analyst-says/).

In addition to sparking enthusiasm and high-content purchases among its loyal customer base, Amazon as a Charismatic Leader is exercising tight control over its Kindle ecosystem. One aspect of Amazon's strategy has been its insistence on a proprietary format for the e-books downloaded on the Kindle. This puts Amazon in the position of gatekeeper for all the Kindle content. Customers can put their own materials onto their Kindle using PDF or text file formats, but content that is sold through Amazon comes in a format that is readable only on the Kindle. Like all Charismatic Leaders, Amazon runs a high risk of competition from more open content and partnering models that are on the horizon.

The Kindle is acting as a catalyst for e-book adoption by consumers but Amazon may not be able to maintain its control over the Charismatic Leader ecosystem model in the face of alternatives that offer more open platforms, advanced design features, or both. In 2010, Sony, Plastic Logic, and other e-reader manufacturers announced competing products. Apple also entered the e-reader sector with the debut of the iPad, its own smart, connected device to compete for a piece of the e-book and media content market. The risk of outside disruption in the absence of constant innovation is one of several factors that make it challenging to succeed over the long run with the Charismatic Leader ecosystem model.

*Charismatic Leader model risks*
The rewards of the successful Charismatic Leader model are high, but so are the risks associated with this strategy. Pitfalls include the constant pressure to maintain a position of advanced technology while developing smarter and smarter product offerings. This is an

important basis for the brand's leadership position and customer loyalty. Even though a Charismatic Leader wants to retain some control over its innovative ecosystem partners, the brand must continually demonstrate that it maintains an edge in innovation and customer adoption. Losing the reputation of industry innovation, or becoming "uncool" can create a downward spiral. Once the aura of charisma is gone, it will be very hard to recreate it. If the brand's products fall behind other market offers or fail to win accolades with each new generation of product releases, the loyalty of consumers may fall off dramatically along with the motivation of partners to invest time and effort in adhering to ecosystem participation requirements.

The constant pressure to generate revenue for ecosystem partners may risk alienating customers who are able to find cheaper devices and often abundant free content in the same product category by looking to alternatives from the Transformer ecosystems. For example, customers who buy a Sony e-book reader can access millions of free Google book titles since Google digital books are formatted in the industry-standard ePub format which is supported by Sony but is not compatible with the Kindle.

Setting a higher and higher bar for smart product and market innovation requires continued investment and an even more difficult challenge of continued smart product design and platform innovation that anticipates and helps to establish new market demand. A slip in the leader's product performance and market acceptance creates vulnerability for the whole ecosystem. As the model name implies, this ecosystem relies on the ability of its central brand to reinvent itself to continually dominate the market. Only a small number of companies can maintain this level of innovation over time.

### Federator model

The Federator ecosystem model is a well-trodden path for companies interested in entering an industry sector and building alliances around new technologies that require the participation of many partners, including existing industry leaders. The Federator can court industry leaders on the basis of establishing shared standards and by undertaking tasks such as technology platform certification that may be outside the current scope of the leading market players.

This strategy can work for new entrants as well as for established industry players as long as they understand the time that it will take to generate support for foundational technology platforms and standards. Microsoft has adopted the Federator model in smart health to present itself as a trusted aggregator and protector of the copious medical and personal health data that is generated from smart, connected health devices. Since launching HealthVault in October 2007, Microsoft has enlisted respected healthcare leaders such as Aetna, CVS, and the Mayo Clinic to link their services to the platform as well as attracting multitudes of application providers and health and fitness device manufacturers (HealthVault, 2009). The HealthVault value proposition for ecosystem partners and consumers alike is the ease of integrating devices and patient records. Since personal medical and health data storage requires strong assurances about the security and confidentiality of the data collected, Microsoft's reputation and stability is also an important factor. For a small company interested in attracting consumers to its smart, connected device, the use of HealthVault as a data platform eliminates the need to set up an expensive separate data storage and access system and makes it easier to convince customers that their information is secure and available for the long term. The existence of Google Health, a similar platform hosted by Google, allows device makers to give consumers a choice between HealthVault, Google Health, and other such services without incurring the considerable privacy and security overhead of storing medical records.

However, the availability of multiple platforms and possibly dueling specifications and standards can become a problem for Federator ecosystem adoption. Forcing partners to pick sides may split an emerging smart product market rather than creating a consensus that encourages new entrants and innovation. Ideally, Federator platforms and formats will be compatible with all others across the industry, allowing ecosystem partners and consumers to use the solution that they are most comfortable with, and willing to trust, without worrying about their records and data being locked out of future migration to other platforms.

Another role that Federators must play is facilitating the interoperability of smart products and services. Another healthcare Federator, the Continua Health Alliance, is set up to provide device certification for its ecosystem partners, a major factor in ensuring smart product interoperability.

> Continua Health Alliance, the non-profit, open-industry alliance of more than 200 leading healthcare and technology organizations, recently announced the availability of three new Continua Certified™ devices from member companies. These certified products mark an important milestone in the group's mission to establish a system of interoperable personal connected health solutions. By extending these solutions into the home, Continua is driving a new model in healthcare, enabling truly personalized health and wellness management that empowers individuals and fosters independence.
>
> *(Continua, 2009)*

Competition among Federator models has been a barrier to adoption in many consumer sectors, including home automation and networked entertainment devices. Similarly, the smart energy sector is crowded with competing technology standards, industry alliance groups, consortia, and federators all contending to become the dominant core of a thriving home energy monitoring ecosystem.

The adoption of the ZigBee wireless standard for connected health devices and smart energy management systems were important milestones for Ember, a Federator which has staked its growth and future profitability on the global adoption of ZigBee chips and smart products. ZigBee, the market name for the IEEE 802.15.4 standard that Ember uses as a core technology for its chips, has the advantage of low power consumption and long battery life. Nevertheless,

ZigBee remained outside of mainstream smart product development until quite recently, illustrating the long timeline required for Federator models to take hold. Ember is an example of the need for deep pockets and a very long-term horizon on adoption for success in the Federator ecosystem model.

Ember has pursued a Federator strategy for almost a decade to become the chip platform of choice for in-building systems automation, connected health devices, and smart energy management. In the process, it has spent at least $89 million in venture capital investment and has yet to achieve profitability. To build the ZigBee ecosystem, Ember staff spearheaded the formation of a ZigBee Alliance, contributed to the evolution of the specification, developed design tools, and recruited industry partners in order to create a market for its chips and development tools; as described on the Ember web page: "Ember has committed significant human, infrastructural and financial resources to help accomplish the mission and goals of the ZigBee Alliance" (Ember, 2009). While Ember seems poised to reap the benefits of its long-term strategy, many new market entrants fail to have the staying power needed to overcome the risks of the Federator model.

*Federator model risks*

Even compared to the long-term perspective required of Transformers, the time commitment required for establishing a successful Federator ecosystem is daunting. This protracted adoption cycle is one of the major risks of the Federator model, especially for small companies. A Federator needs to have significant funding available in order to promote its technology platform and related standards and recruit new partners to join the ecosystem well in advance of any significant revenues. Small players risk making this multi-year investment only to lose momentum before it begins to pay off.

Other risks are that a Federator may fail to convince industry leaders to join the ecosystem or end up competing with potential partners who decide to set up a competitive model based on a different standard. Many new industry sectors suffer from an over-abundance of aspiring Federators who are all claiming the banner of platform standardization and interoperability. This competition can add to the time it takes for partners to adopt any platform and may confuse customers rather than reassure them about the stability and long-term future of the Federator-affiliated smart products and services.

To mitigate these risks, many Federators ally closely with standards organizations or start their own industry alliances to gain consensus for shared ecosystem specifications and encourage early adoption by other alliance members.

### ROLE OF THE CUSTOMER

The choices that companies make about their ecosystem strategy set the parameters for the smart product value proposition, extent of control over ecosystem partners, and the expected activities of customers using smart products developed and supported by the ecosystem. Each ecosystem creates implicit or explicit structures that guide customer expectations and behaviors in interacting with the product, defining the various roles that the customer is invited to play, or limited to playing, within the ecosystem. These structures create a default mode for a customer's ecosystem participation.

One characteristic of Hegemon models is the tight control of smart product features and use cases, making it difficult for customers to experiment with alternative uses or adopt substitute products for any component of the ecosystem. Hegemon customers are expected to remain inside the ecosystem and rely on authorized partners for all their product and service needs. The default customer role is citizenship. Customer opinions are welcome, as are the modes of participation laid out for good citizens inside the ecosystem. If customers want alternatives, their primary recourse is to vote with their feet; that is to leave the ecosystem to find an alternative product or service. Since Hegemon companies are often the dominant – and sometimes the only – provider of a particular product, they don't risk erosion of their customer base unless new entrants succeed in disrupting their model with alternative product offers.

Since the Charismatic Leader typically already has a large group of customers, it rewards loyal followers who come along to the new industry sector with innovative smart services and product choices. Customers tuned to keeping up with trends and staying ahead of the mass market see themselves as early adopters and trendsetters when they join a Charismatic Leader ecosystem, roles that they enjoy playing. At the same time, this brand-dominated ecosystem minimizes the risks and frequent frustrations of early technology adoption in more open models. In the case of iPhone and Kindle, customers are already familiar with the brand's online user interface and content distribution system, so they have a head start on optimizing their use of the new devices, another reward for loyalty. The Charismatic Leader strives to make its customers into enthusiastic fans and to turn these fans into avid product evangelists. Without the perspective of a fan, it is not as likely that customers will remain content with the brand's control as the market matures and more open, lower-priced smart product alternatives become widely available.

The ideal of the Federator model is to place the customer in control of a simple interface that seamlessly supports every component and product offered by its ecosystem partners. Federator models that use standards to enable multiple manufacturers to connect their products emphasize this variety of choice and availability of products that all work on the same standard. However, the Federator may leave it to the individual customer to figure out how to make all these different products interoperate, with frustrating consequences for consumers. This means that the target customers must often take a more active role in managing their own products and services. The Federator's default customer role is as a manager or a team leader. If product interoperability hasn't achieved a plug-in level, customers may have to act more like general contractors; a role that far fewer smart product buyers are willing or able to fulfill.

The Transformer model also relies on customers who are willing to engage actively with cutting-edge products and services, especially early product versions that may still have some kinks to work out. These customers are likely to enjoy being on the leading edge of successive generations of innovative technology. They are willing to take the risk that their early product purchases may be discontinued or quickly outdated as new products are developed or, worst case, that the Transformer fails altogether. Such risk-takers are motivated to explore the product's features and are typically interested in providing feedback to the vendor as well as exchanging tips and solutions with other early adopters. They prefer being able to choose from a variety of brands and service providers rather than being locked into one dominant brand's product family.

The Transformer consumer may also be motivated by the economic advantages of an ecosystem that provides many different pricing, support, and distribution options, especially as the ecosystem expands. The Transformer customer is likely to be willing and even happy to put in considerable time with the product to understand and improve how it works, making the open platform and availability of developer support and detailed product information an added value. Transformer customer roles include the explorer and the optimizer.

CONCLUSION

In evaluating the pros and cons of different ecosystem models for smart products, companies need to consider a number of different factors. The framework for strategic ecosystem formation presented in this chapter posits two primary dimensions that can influence a choice of ecosystem strategy, including a desire for control vs. openness and disruptive vs. incremental approaches. The benefits and risks of each dimension will apply differently to different industry sectors and product categories as well as to each company's market position. Decisions that must be made include the added value that each model offers to a company's current and prospective ecosystem partners and the opportunities for additional growth and sustainable advantage in priority market sectors.

Connected smart products provide a dynamic, scalable platform for developing and delivering smart services. Smart ecosystems may also be deployed as mechanisms to control and restrict customer choices and modes of interaction with the product. The critical choice of an ecosystem strategy sets a balance of control and openness that may hinder or facilitate the consumer adoption of smart services. Many companies gravitate toward ecosystem control features as a defense against commoditization of their products and the escalation of digital piracy. However, if the preponderance of customer interactions in a smart ecosystem become control-based then the ecosystem may be less effective in attracting and retaining customers and generating recurring revenues. The roles that it allows its customers to play in the ecosystem may be too restrictive to promote innovation and continuous technology and service development. Chapter 3 analyzes the ways in which smart products are utilized for control and contrasts an emphasis on embedded control with the use of smart ecosystems for the creation of high-value service opportunities.

# 3 Embedded product controls

Tom finally has the family car keys in his pocket. He has waited a long time for this moment, having suffered through driver education and envied his older friends who already have their own cars. Tom's first solo drive won't be quite as exciting as he would like, however. His new set of keys are more than a ticket to ride; they are microcontroller-enabled smart keys linked to Ford Motors' MyKey driver control system, with safety features designed to keep Tom within limits.

The MyKey system allows Tom's parents to set the maximum speed that he can drive their Ford Focus: when he puts the key in the ignition the car reads his personalized driving profile and sets its controls accordingly. To help ensure that Tom keeps his eye on the speedometer and his foot off the gas pedal, a reminder chime will sound when the car reaches preset speed levels of 45, 55, and 65 miles per hour (MPH). The system takes control even before Tom leaves the driveway, enforcing his use of seat belts by sounding a recurring reminder chime and keeping the audio system muted until the driver's belt is buckled.

Tom isn't exactly happy with all these limits on his independent driving. But the alternative would be much worse – waiting until after high school graduation before getting access to the family car. That delay was a possibility after his friend Jack took an unfamiliar route home from a party, missed a turn and ended up in a ditch, wrecking his car in the process. The accident investigation revealed that Jack was speeding, not wearing a seat belt, and not in control of the vehicle when it left the road. Luckily, he wasn't badly hurt but it was his second crash in three months and that was enough to get

Jack grounded and to convince Tom's parents to keep close tabs on their son's driving.

Cars that control driver options aren't news to Tom. He knows that his father puts up with a much stricter driver monitoring program on the job; every employee car in his father's company is equipped with a system that records vehicle speed, braking, steering, and acceleration patterns, reporting on any behavior that indicates unsafe driving habits. In order to reduce the cost of auto insurance, his mother is considering a pay-as-you-drive program with Progressive Insurance that will mean installing a mileage monitor in her car. And some of his friends have it even worse. Their parents have installed GPS monitoring systems to track and report on every place their children drive.

Tom is learning fast that these days the key to the family car comes with a lot of restrictions. In fact, Tom and his friends have grown up in a world where usage restrictions and outside monitoring are facts of everyday life. Every digital versatile disc (DVD) they watch comes with a formidable warning from the FBI prohibiting copying and threatening "severe civil and criminal penalties" for unauthorized reproduction, distribution, or exhibition. Every piece of software they install and each new online service that they sign up for require them to agree to pages of complicated legal language and licensing terminology. This generation of adolescents may soon view the automobile as just one more product that enforces rules set by someone besides themselves.

Is MyKey an example of embedded product control or is it a customer-selected option? Like many of today's smart, connected products, MyKey has aspects of both. Smart auto systems with embedded controls may attract customers with a specific need to monitor the behavior of other drivers. To these customers, driver control is a valued service. Worried parents of teen drivers and managers responsible for reducing the accident rate of their fleet vehicles are good matches for the value proposition of products and services that control driver behavior.

Auto insurance companies looking for innovative ways to manage the risk associated with issuing policies to potentially crash-prone drivers are also attracted to in-depth driver monitoring technologies.

However, embedded controls may be resented by consumers who have to submit to them. If the individual driver's choices behind the wheel become too constrained, that can detract from the positive aspects of owning and driving a car. Drivers who chafe under strict embedded controls may well be less likely to identify with the automobile as a source of personal fulfillment. Customers may abandon their brand identification and loyalty to overly controlled products and the entire product category perception may shift. According to some reports, that shift in perception is already happening in Japan, where the next generation of young drivers is less interested in buying a car than in Internet, music, and smartphone services. In a trend that worries Japanese auto executives, consumers in their twenties "see cars as nothing more than a tool, much like a vacuum cleaner, not a reflection of their identity, tastes or income level" (Asia Finest Discussion Forum, 2009).

The framework for smart ecosystem strategies presented in Chapter 2 addressed the benefits and risks of four different ecosystem models from the perspective of the ecosystem brand and its business partners. Selecting the most appropriate ecosystem model is just one element of a successful smart product strategy. Equally important is creating a balance between exercising embedded product control and using embedded intelligence and connectivity to deliver a service that is valued by customers. This chapter discusses the technical, regulatory, and legal avenues that product vendors and service providers are using to exercise control over smart products. In the process, the chapter will analyze the business models that exploit smart product capabilities exclusively for the benefit of the vendor and the impact of more aggressive consumer control strategies on a smart product's value proposition. Subsequent chapters will present examples of smart product services that provide alternatives to primarily control-based strategies.

## A TYPOLOGY OF CONTROL

There are a variety of options for a smart product to exercise control over the way it is used by its owner. Table 3.1 categorizes and defines four common modes of control in smart product design and deployment.

Many vendors incorporate a combination of the approaches in Table 3.1 to maximize their ability to control consumer use of their products. These modes of control are not unique to smart products. Terms of use and contracts of various kinds have long been employed to establish the expected parameters of product use. Protective controls such as safety locks and product interlocks have existed for many years and may be as simple as an appliance door that is locked when the appliance is in operation. The OnStar service described in Chapter 2 is an instance of consumer-delegated control in that the driver expects OnStar to automatically notify the police and summon help if the car has been in a serious accident. As the following sections will discuss, smart product features can be deployed to exercise pre-emptive vendor control, protective control, and consumer-delegated control depending on the context. There is also a trend to rely on pre-emptive control to enforce consumer compliance with a variety of product restrictions that are included in Terms of Use (TOU) and End-User License Agreements (EULA).

### Delegated consumer control vs. pre-emptive vendor control

In many cases, product owners see a value in delegating control to their smart products. They may even pay a premium for the embedded control option. Ford's MyKey is a case in point. Parents purchase the MyKey system to ensure that their children's driving stays within specific limits. They gain peace of mind from delegating control over speed and safety belt use to MyKey. Car owners may also delegate control of their car to smart driver assistance features such as automatic cruise control or collision avoidance systems. In a home with smart energy consumption systems, consumers may delegate control over the heating and cooling units to their utility company. The

Table 3.1 *Smart product modes of control*

| Mode of control | Definition |
| --- | --- |
| Pre-emptive control by vendor | Vendor defines the limits of product functionality and imposes controls on usage in advance of sale through embedded programs or post sale by updating the product parameters without the explicit agreement or even knowledge of the owner. Product owner is not able to override the limits imposed by the vendor. |
| Delegated control by consumer | Product owner agrees to the automatic functioning of a product on their behalf, either using the built-in product parameters or proactively requesting the vendor to control the product on their behalf. |
| Protective controls including public sector regulations and private sector health and safety options | Built-in product usage restrictions that are explicitly designed to prevent injury to the individual using the product or to others who come in contact with the product. |
| Compliance and contractual controls | Product usage restrictions that are defined by Terms of Use, End-User License Agreements or other types of contracts that create legal liability for owners who violate the agreement. May include embedded usage restrictions that embody the vendor's interpretation of copyright and other legal protections for product content and software. |

consumer makes an agreement with the utility company to install a smart thermostat that is connected to a smart electric meter. In exchange for lower electricity rates, the consumer agrees that the utility company can turn off or cycle down air conditioning and other connected appliances to reduce the strain on the electricity grid during periods of peak demand.

Delegated control is a choice the consumer makes, and the delegation can be withdrawn in the future if the product owner desires. Parents can stop using the MyKey system once their teen driver has established a safe driving record. Consumers can opt out of the utility company control of the thermostat if the arrangement turns out to be inconvenient.

Pre-emptive control is imposed by the product vendor, often without the knowledge of the consumer who has no choice to opt out. The GPS and telematics technology that enables vehicle safety systems such as OnStar to connect remotely to car locks, ignition, and engines, including a service that allows police to recover stolen cars by literally stopping the thieves in their tracks, can in a different situation be programmed to override consumer preferences through monitoring and then take control from a distance. Another vehicle system with embedded intelligence and connectivity, dubbed the Disabler, is not likely to be welcomed by car owners. The Disabler has been used by car dealers to keep track of customers who purchase their vehicles on a payment plan. If the owner falls behind on the car payments, the dealer can use the Disabler to prevent the vehicle from starting while pinpointing its location for a quick repossession (Welsh, 2009).

Locking up and disabling mobile phones is a less dramatic but equally complete shift of product control away from the owner and back to the wireless carrier or device maker. Wireless carriers routinely lock the subsidized handsets they sell so that the handset will work only with that carrier's subscriber identity modules (SIMs) and thus with the account they have with the customer. This prevents customers from buying a phone from one carrier and using the phone with a subscription from another carrier. Carriers can enable their subscribers to unlock their handsets simply by providing an authorization code that the subscriber enters on the phone's keypad. Unauthorized unlocking attempts are common, but they come with a significant downside for the subscriber. If the consumer enters the wrong code, the handset is programmed to lock up permanently after a predesignated number of wrong entries, usually between three and five. The only recourse for a subscriber with a disabled carrier-locked handset is to go back to the carrier and have it re-enabled. Unauthorized unlocking attempts are often cited in the mobile phone warranty or the subscriber's terms of service as an action that will void the handset warranty and terminate any future handset support from the carrier.

In order to activate Apple's iPhone in the USA, consumers have to sign up for two years of wireless service with AT&T. Customers who bought iPhones with the intention of unlocking the link to AT&T and bypassing embedded iPhone restrictions were given a clear warning about pre-emptive vendor control of smartphones. Instead of an iPhone that would work on any network, subscribers who tried to unlock their iPhones would risk ending up with a disabled device and voided warranty. In an official statement in September 2007, Apple announced to iPhone owners that any attempts to unlock their phones or add third-party content or software from sources outside the Apple App Store was a violation of the terms of the Apple software license agreement and as such would void their warranty, saying in part:

> many of the unauthorized iPhone unlocking programs available
> on the Internet cause irreparable damage to the iPhone's
> software, which will likely result in the modified iPhone
> becoming permanently inoperable when a future Apple-supplied
> iPhone software update is installed ... Apple strongly discourages
> users from installing unauthorized unlocking programs on their
> iPhones. Users who make unauthorized modifications to the

software on their iPhone violate their iPhone software license
agreement and void their warranty. The permanent inability
to use an iPhone due to installing unlocking software is not
covered under the iPhone's warranty.

(MacDailyNews)

Locking subsidized mobile phones and restricting the functionality of selected aspects of the phones both before and after sale
are long-standing practices of wireless carriers who operate within
a centrally controlled Hegemon ecosystem. Since the mobile device
and all its network features must be activated on the carrier network
and since the device is connected to the network whenever the owner
wants to use it, the carrier can exercise control over everything from
the handling of voice and data communications, to access to non-cellular communication channels such as Wi-Fi, Bluetooth or GPS, to
the management of applications and content resident on the device.
When consumers attempt to work around the carrier's pre-emptive
controls, it is a simple matter for the carrier to access the mobile
phone and, for example, delete any unauthorized features. The network connection allows carriers to carry out the ultimate sanction
of disabling the SIM and the phone entirely. Even this drastic control
can also be used as a service in other circumstances, however. If a
subscriber's mobile phone is lost or stolen, it is to the subscriber's
benefit when the carrier disables that phone remotely.

With the proliferation of connected smart devices, the exercise of pre-emptive control becomes available to many more product vendors. In his discussion of the Internet's role in controlling
customers and their smart products, Jonathan Zittrain calls these
connected products "tethered appliances" because they continue to
be tied to their manufacturer or vendor and subject to revision even
after the consumer has taken possession. Unlike traditional physical
products that will remain in the same state as they were on the day
of purchase, tethered devices can be polled, updated, and disabled by
a remote command long after a consumer has paid for and become

dependent on them. According to Zittrain, the features of tethered appliances are contingent upon the will of the vendor, placing them in a new class of technology (Zittrain, 2008).

If we like our connected game console, or our phone, or our e-book reader exactly the way it is today, can we do something to make sure it doesn't change behind our backs into a product that acts differently? Unless a consumer can manage to turn off all connectivity between the product and its network (a step that may disable one or more critical aspects of the product's functionality) the short answer is no, the owner does not have the ability to freeze and continue to use the current version of their smart products if the vendor is determined to update them. The smart product TOU and license agreements typically give the manufacturer or the vendor, or both, the right to update the basic operations, content, and interoperability options of the smart product in whatever way – and whenever – they want to. And if the vendor has already embedded a program telling the product to follow authorized vendor instructions rather than owner instructions in certain cases, it doesn't need consumer agreement every time something is updated or deleted. The vendor has reserved the right to exercise pre-emptive control.

This is a lesson that consumers learn over and over again as each new smart product platform that they adopt asserts its control. Amazon's unannounced deletion of e-books by George Orwell in the summer of 2009 was a surprise to the Kindle customers who had bought and paid for the affected titles directly from Amazon. One minute the Orwell e-book was on their Kindle, the next minute it was gone without a trace. This deletion was a wake-up call for all Kindle owners that Amazon retains control of the contents of the Kindle even after a book has been purchased, downloaded, and personalized with notes and annotations. Amazon provided the affected Orwell readers with a credit and later stated that it would not reach out and delete content in the future. But Kindle's embedded wireless connection to Amazon, a highly valued feature of the device for many customers, had demonstrated the darker side of remote control access by vendors. The fact that the deleted works were books by George Orwell, whose book *1984* is synonymous with authoritarian control and loss of personal freedom, added irony to this example of pre-emptive smart product control.

*Protective controls*
Safety hazards in mechanical equipment, the resulting injuries to equipment operators, and government safety regulations provided the original impetus for designing a form of protective control called product interlocks. Interlocks were originally mechanical and were intended to protect operators of machinery, for example industrial press machines or power saws, from injuring themselves in the process of using the device. A power saw guard, for example, would shut off power to the blade if a block of wood or a hand was inserted inside the housing protected by an interlock. Without this control, machine operators could – and often did – lose their fingers when reaching too close to the blades. Many interlocks of this type have been mandated by government and employee safety regulations and product safety codes (Lockton, 2005).

Safety interlocks are common on household appliances as well as industrial machinery. Household interlocks, for example, ensure that the doors of microwaves, washing machines, or self-cleaning ovens are not opened during operation. An interlock can be something as simple as a door you have to lift before you flip a switch or as complex as a school bus component that senses a child in front of the bus and puts on the brakes. It can be easy to override, or completely impervious to consumer intervention, depending on its purpose and design. Interlocks exercise a form of pre-emptive consumer control, but most consumers willingly accept these constraints as a trade-off for the greater good of health and safety protection. In fact, consumers who have been injured using a product as directed have brought lawsuits and been awarded compensation from the manufacturer when it is determined that the injury could have been prevented by better safety design, including use of product interlocks (Leveson and Turner, 1993: 18–41).

Like many forms of product control, product interlocks now include connected embedded intelligence, allowing them to exercise control over a broader range of consumer behavior. In the process, the acceptance of interlocks as a clear-cut consumer benefit has been eroded. For example, interlocks that prevent drivers from starting their cars after drinking alcohol are controversial, even though these locks have been adopted by many US states and several countries in an effort to save lives and improve public safety.

Alcohol ignition locks combine an interlocking component on the automobile ignition with sensors that measure the level of alcohol that has been consumed by the prospective driver. One vendor defined these products as follows: "An ignition interlock device is an in-car alcohol breath screening device that prevents a vehicle from starting if it detects a blood alcohol concentration (BAC) over a preset limit" (Smart Start, 2009). To prevent an inebriated person from asking someone else to start their car and then taking over the wheel, the alcohol interlock may prompt the driver to breathe into the device at regular intervals while driving.

Mothers Against Drunk Driving (MADD) advocates that every state in the USA require judges to order the installation of ignition locks in the automobile of anyone convicted of driving under the influence of alcohol (DUI). MADD points out that drunk driving was implicated in thousands of highway accidents and over 11,000 fatalities in the USA in 2008 and notes that many convicted drunk drivers are repeat offenders (Mothers Against Drunk Driving, 2009). The MADD position is that preventing such drivers from even starting their cars when intoxicated is the only effective method of reducing alcohol accidents and fatalities. More than forty states in the USA and a number of other countries have come to the same conclusion, and now require drivers convicted of repeated DUI offences to install an ignition interlock in their car as a condition of getting back behind the wheel. Currently over ten US states require installation of interlocks even for first DUI offenders. Some safety advocates go even further and recommend the installation of alcohol-sensitive interlocks in every vehicle, so that no one with an elevated blood alcohol level will be able to start their car or drive it even if they have never had a DUI conviction.

Factory-embedded interlocks with the Volvo brand name of alcolocks are already available on selected Volvo cars and trucks. Sweden is considering legislation to make alcohol interlocks mandatory on all vehicles driven by public employees, including buses and trucks. Instead of the cumbersome breathalyzer style of interlock that requires drivers to blow into a tube, the Volvo alcolock uses sensors on the driver's seat belt to register traces of alcohol. Volvo is in favor of mandating such alcolocks on commercial and public vehicles, stating on its website: "Volvo is a strong advocate of the introduction of the alcolock in every commercial vehicle. The technology is available and the cost is becoming less" (Volvo Technology Magazine, 2008).

Predictably, one staunch opponent of increasing the deployment of alcohol interlocks in the USA is the American Beverage Institute (ABI), the industry association which represents restaurants and bars. The ABI came out fighting against federal legislation that would require all states to mandate interlocks for first-time DUI offenders in a June 2009 press release, stating: "Today the American Beverage Institute (ABI), which represents thousands of American restaurants, denounced a component of the Highway Bill which would require that all 50 states mandate ignition interlocks as punishment for low-BAC (blood alcohol concentration) first-time offenders" (American Beverage Institute, 2009).

ABI also raises the specter of alcohol interlocks malfunctioning and stranding hapless motorists, even in cases where no alcohol has been consumed, requiring car owners to call for costly roadside assistance simply to start their cars. The ABI accuses the auto industry of allying with MADD and government regulators to support alcohol interlocks and predicts that auto makers will profit from interlocks first by adding the extra cost onto the sticker price of the car and then by servicing the inevitable malfunctions that such interlocks will be subject to during the lifetime of the vehicle. As with many regulated embedded controls, there may be legal penalties for consumers who disable a factory-installed ignition lock even if it is malfunctioning or hampering the legal operation of the vehicle. Anti-tampering regulations and the possibility of owners incurring penalties for disabling an embedded control for reasons that are not themselves illegal are not unprecedented; they are already in place in the provisions of the Digital Millennium Copyright Act, which we will discuss in more detail later in this chapter.

## PROPRIETARY APPLICATIONS FOR EMBEDDED CONTROLS

It is a very small step-up for manufacturers to shift from installing embedded controls that protect consumer safety to extending these controls to proprietary strategies for business advantage. The evolution of On Board Diagnostic (OBD) systems in automobiles from a tool for measuring vehicle emissions to an embedded smart technology platform that monitors engine performance is an example of how manufacturers can use embedded smart product components to extend their sphere of control over an entire ecosystem. In the case of OBD modules, this ecosystem includes auto component makers, car dealerships, independent service garages, mechanics, and consumers.

The first OBD standards (now dubbed OBD I) were developed by the California Air Resources Board in 1988 as part of the push to decrease pollution from automotive emissions and improve air quality. The specific capabilities of the next generation of OBD (OBD II) were spelled out in the federal Clean Air Act Amendments of 1990 with a mandate that cars and light trucks sold in the USA from 1996 forward must be equipped with an OBD II system. According to the Environmental Protection Agency (EPA) website:

The OBD II system monitors virtually every component that can affect the emission performance of the vehicle to ensure that the vehicle remains as clean as possible over its entire life, and assists repair technicians in diagnosing and fixing problems with the computerized engine controls. If a problem is detected, the OBD II system illuminates a warning lamp on the vehicle instrument panel to alert the driver. This warning lamp typically contains the phrase Check Engine or Service Engine Soon. The system will also store important information about the detected malfunction so that a repair technician can accurately find and fix the problem.

*(Environmental Protection Agency, 2009a)*

In a note that echoes concerns about potential non-tampering provisions that might come along with any mandated alcohol interlocks in future car models, the EPA website also cautions consumers to keep their hands off the OBD system:

The rule of thumb when it comes to emissions-related vehicle repair is that any modification that changes the vehicle from a certified configuration to a non-certified configuration is considered tampering: this applies to both vehicle owners and repair facilities and is, therefore, a Federal offense ... Likewise, overriding the OBD system through the use of high-tech defeat devices or non-certified computer chips, for example, would also be considered tampering.

*(Environmental Protection Agency, 2009b)*

Faced with a mandate to build in an expensive and complex diagnostic system to monitor and record the performance of the many components involved in creating automotive emissions, auto manufacturers looked for ways to leverage their investment in the OBD platform. An obvious answer was to make the OBD a more general-purpose diagnostic tool for all types of vehicle performance and to extend it to support vehicle-specific and manufacturer-proprietary performance-monitoring features. Since the EPA regulations left manufacturers free to add their own proprietary diagnostic codes as long as their OBD II systems also met the standard EPA requirements, some auto makers added hundreds of different sensors and diagnostic codes to each car's diagnostic system.

The proliferation of proprietary diagnostic codes available inside extended OBD systems created a need for special diagnostic software and equipment to read out all the data and organize it into information formats that would be useful to mechanics. Auto makers and authorized ecosystem partners produced diagnostic kits that mechanics could use to download relevant engine data and create performance issue reports when customers brought their cars in for servicing. As more proprietary OBD extensions were developed, the specialized engine diagnostic software and readout tools became a competitive advantage for car dealerships, since they had the best access to all the required tools for new car models.

Each car maker has a somewhat different set of diagnostic codes and software for reading out the OBD module at the repair bay. Some of these codes have been standardized but hundreds of proprietary codes still exist and are updated continually by the car manufacturers. This makes it essential for mechanics to buy the proprietary software programs that can interpret the output from each make, model, and year for each brand of vehicle. Keeping up isn't a problem for the authorized dealership garages where only one or two different car makers are serviced and where the auto makers will provide the software under the franchise agreement. However, this type of software and diagnostic equipment limits the choice of mechanic for the car owner and puts a damper on all but the most trivial do-it-yourself repairs. It also serves as a de facto barrier for ecosystem participation by small garages and independent mechanics.

A search of OBD-Codes.com, one of the many online sites that provide a listing of diagnostic vehicle codes, underscores the difficulty of keeping up with these proprietary vehicle code systems. The list of most common generic OBD codes used by multiple auto makers has more than 500 entries; among fifteen different vehicle brands listed on the website are thousands more codes that are specific to vehicle year, model, and body type. These free listings are only a starting point for anyone who wants to repair a car. As the website notes, they are not comprehensive or updated regularly enough for mechanics to rely on them:

> This list contains standard diagnostic trouble codes (DTC's) [sic.] that are used by some manufacturers to identify vehicle problems. The codes provided below are generic codes that may not apply to all vehicles. Vehicle manufacturers may use manufacturer specific DTC codes that are different from the codes shown below. Foreign vehicles may also use DTC codes different from the generic DTC codes. We recommend that you not depend on the DTC codes in this list for vehicle repairs until you confirm the generic DTC codes apply to your vehicle(s). The list below is for information purposes only and is not intended for use in vehicle repairs.
>
> *(OBD-Codes.com, 2009)*

Years of legal battles by independent mechanics have resulted in broader availability of the diagnostic software and equipment, but the cost of maintaining the latest information for every make and model remains an economic burden for smaller garages trying to provide repairs on a broad variety of newer vehicles. The original government mandate for an embedded engine diagnostic component for standardized measurement of vehicle emissions has become a complex and sophisticated system for auto makers to monitor and log the performance of hundreds of components and to extend their control over vehicle repairs long after the vehicle leaves the dealer showroom.

Companies can leverage embedded intelligence to exercise ecosystem control even without regulatory mandates. But the combination of legal restriction with embedded pre-emptive controls extends the vendor's sphere of control and options for enforcement.

Companies have long relied on patenting their products and technologies to prevent competitors from offering comparable solutions, thus indirectly limiting the options available to consumers. Connected smart products provide an even more direct avenue to control, especially when backed by the force of law. Once legal protection is granted to one market sector, it can be adopted by companies in other sectors to impose even further controls that may seem far afield from the purpose of the original legislation. This has happened with the Digital Millennium Copyright Act (DMCA), which was originally focused on protecting digital media. The DMCA has subsequently been invoked by companies making items as diverse as garage door openers and printer cartridges against competitors who are building products with the same functionality, or providing generic replacement parts for peripherals within a controlled product ecosystem (Electronic Frontier Foundation, 2010). As more companies have adopted strategies for restricting consumer behavior rather than simply using patents to create barriers to entry for competitors, the EULA and TOU associated with smart products have taken on new significance. The implications of combining pre-emptive control with legal and contractual controls to govern consumer use of smart products are analyzed in the next section.

## PATENTS AS BARRIERS TO ENTRY AND EMBEDDED CONTROLS AS BARRIERS TO EXIT

Patents have long been a cornerstone of competitive strategy, frequently used to buttress the links between proprietary products, peripherals, and replacement parts. Gillette's well-known approach of offering low-cost razors and generating profits from selling disposable razor blades was based on multiple patents that kept competitors from making compatible blades. More recently, Nestlé has created a line of espresso machines called Nespresso™ that relies on a patent wall to protect its highly profitable sales of proprietary coffee capsules. Nestlé doesn't manufacture the coffee makers. They license others to build Nespresso machines and sell them through retail stores worldwide. Nestlé controls the market for distributing the consumable component of the product, the little capsule filled with coffee that is used in a Nespresso machine to make a beverage. These capsules cost about fifty cents each and must be purchased in quantities of at least fifty at a time.

The capsules are little more than thin-walled sealed containers that are neither functionally complex nor hard to manufacture and fill. Nonetheless, Nestlé, through its internal R&D company Nestec, has applied for – and been issued – an ongoing series of patents starting in the late 1970s on the Nespresso machines, the disposable capsules, and virtually all aspects of the process of poking a hole in the capsule and forcing water through it. Selected patents related to Nespresso capsules are listed in Table 3.2.

Table 3.2 *Selected patents covering Nestlé's Nespresso machine and its capsules*

| US patent number | Title | Priority date |
|---|---|---|
| 4,136,202 | Capsule for beverage preparation | December 17, 1976 |
| 4,724,752 | Machine for the automatic preparation of a beverage | April 30, 1985 |
| 4,744,491 | Device for the distribution of cartridges in a cartridge carrier for a coffee machine | June 28, 1985 |
| 4,806,375 | Method for the extraction of sealed cartridges for the preparation of beverages | June 27, 1986 |
| 4,846,052 | Device for the extraction of cartridges | April 24, 1986 |
| 5,008,013 | Filter for use in a cartridge for the preparation of a beverage | October 1, 1987 |
| 5,151,199 | Process and apparatus for controlling froth of an instant beverage | July 21, 1989 |

Table 3.2 (*cont.*)

| US patent number | Title | Priority date |
|---|---|---|
| 5,197,374 | Apparatus for extracting cartridges containing coffee | July 27, 1990 |
| 5,242,702 | Extraction of coffee contained in sealed cartridges | July 27, 1990 |
| 5,327,815 | Device for use in beverage extraction machines | July 5, 1991 |
| 5,347,916 | Device to assist extraction of beverage material in cartridges | July 27, 1990 |
| 5,398,595 | Devices for extracting beverage materials in cartridges | July 5, 1991 |
| 5,398,596 | Extraction device for preparation of a beverage | July 27, 1990 |
| 5,402,707 | Obtaining fluid comestibles from substances contained in cartridges | May 8, 1991 |
| 5,637,335 | Capsule package containing roast and ground coffee | July 5, 1991 |
| 5,897,899 | Cartridges containing substances for beverage preparation | May 8, 1991 |
| 5,948,455 | Cartridge having shared thinned areas for promoting opening for beverage extraction | May 10, 1996 |
| 6,026,732 | Holder and housing devices for containing and releasing capsules for drink preparation | July 14, 1997 |
| 6,849,285 | Sealed cartridge for making a beverage | March 18, 1999 |
| 6,880,454 | Capsule cage | June 30, 2000 |
| 6,832,542 | Method and device for preparing a hot beverage | September 20, 2002 |

Nestlé uses its extensive patent portfolio to capture all the Nespresso capsule sales by the simple expedient of being the only entity legally entitled to make compatible capsules. The patents ensure that Nestlé controls the market for Nespresso capsules.

Nestlé could use this control to raise the price of a Nespresso capsule from fifty cents to $2.50 each. Or they could increase the minimum purchase to 500 capsules per order. Of course customers don't have to buy the capsules under these new conditions but the expensive Nespresso machine that the customer has already purchased then loses its value as a coffee maker because competitors are not allowed to make capsules that will work with it. The fact is that as long as Nestlé controls the market for Nespresso capsules, they control the customer's use of the Nespresso machine too. However, Nestlé has not used its ecosystem control strategy to charge customers the highest possible prices. Instead, it has focused on promoting Nespresso as a premium and convenient coffee-making experience to justify the relatively high price of the proprietary capsules.

This strategy has been a success for Nespresso, making it the fastest-growing unit of Nestlé since its introduction in 1986 with cumulative sales of over 10 million coffee makers and revenue of 2 billion Swiss francs in 2008 (Nespresso, 2009). Its formidable array of patents has kept competitors from offering generic Nespresso capsules on the open market. Nespresso didn't build this success around pre-emptive control and monitoring of its individual customers. The controls are embedded in the linkage between coffee maker and capsule design and its ecosystem is protected by patents and business barriers.

There are numerous posts on the Internet and an instructive YouTube video with advice on how consumers can make their own Nespresso-compatible capsules. It's unlikely that the customers who post such instructions and videos will ever be sued by Nestlé for encouraging circumvention of embedded controls. It is even more unlikely that customers who decide to try making their own refillable Nespresso capsules at home will be prosecuted. Since the patented coffee capsule is a solidly physical object rather than an easily copied piece of digital software or content, consumer experimentation in refilling the capsules or producing their own home-made versions is no real threat to the profitable Nespresso capsule restocking business. Any company making Nespresso clone capsules for resale, however, is highly likely to be taken to court. The patent wall was built to keep competitors out of the market and to prevent them from making clones of the consumable at a cheaper price point. It has succeeded admirably.

Rigorously enforced patents provide an effective barrier to entry and a foundation for extended product ecosystems, but they have a number of drawbacks. It is time consuming and very expensive for companies to develop a portfolio of patents to protect all the critical aspects of a new smart product and smart ecosystem. Once a patent application is filed, the patent typically takes several years to issue and there is always a risk that issuance will be denied by the patent examiner. Applications must include specific details of the technology innovations, and these details will be published by the patent office, making them available to anyone interested in copying the technology. To keep competitors from building clones, patent holders must police possible violations and be willing to take violators to court if necessary. After all this time and expense, the duration of patent protection is relatively short. In the USA patent protection extends for twenty years for utility patents and only fourteen years for design patents.

The combination of embedded product intelligence, connectivity, and legal restrictions on customer behavior provides a very cost-effective alternative to patent protection. It also allows companies to create strong barriers to a customer exiting the ecosystem since smart products may be designed to cease operating if the consumer stops paying subscription fees or to shut down if the consumer attempts to use the protected product with third-party components or content. A locked smartphone will not operate with another carrier's SIM, for example, and a Kindle will not display an e-book downloaded free from Google. This shift from barriers to competitive entrance to barriers to customer exit has many advantages for vendors, but it creates little value for the consumer.

Unlike patents, contractual and legal restrictions on tampering with or cloning a device, or copying its contents, can be applied as soon as new products are released. Instead of having to pay to develop innovative technology and file patents, companies can simply claim protection for basic product components under copyright and terms of use provisions. Competitors are explicitly prohibited for reverse engineering controls that are protected by the provisions of the DMCA. Unlike patents, which offer only the protections that have been granted in the patent claims, the embedded controls and terms of use for smart products can be updated as often as necessary to keep pace with the availability of new control technologies and competitive products. If the software components of these controls are copyrighted as original creative works, the protection lasts much longer than that of a patent; in the USA copyrights are in force for the lifetime of the creator plus fifty years.

It is not surprising then, that so many companies are shifting their ecosystem protection strategies from creating barriers that prevent the market entry of competitors to creating barriers to their customers exiting the ecosystem. In the age of connected smart products, controlling customer behavior may seem like the fastest, cheapest, and most direct path to protecting revenues and increasing profitability. The provisions of current copyright law allow vendors to prosecute their customers for even attempting to circumvent embedded controls in smart products. Though the original justification for the DMCA was preventing illegal distribution of digital media through the Internet, its restrictions have been applied to all types of products during the past decade in a combination of contractual plus pre-emptive control (Electronic Frontier Foundation, 2010).

PRE-EMPTIVE CONTROL PLUS CONTRACTUAL CONTROL

By the late 1990s the rapid growth of the World Wide Web was seen as a threat to the traditional business models of market leaders in many industries, in particular those sectors that sold and licensed digital content. Internet users were enthusiastic, insatiable consumers of all forms of digital media, and had proven to be extremely ingenious in finding ways to access digital content without paying for it. The rise of rampant copying, sharing, downloading, and stealing of all types of previously profitable and protected content owned by broadcast, entertainment, and media companies prompted those industries to petition for more effective government regulations to protect the copyrights of digital media. The ease of cloning and distributing digital media an unlimited number of times without degrading its quality created support for recasting copyright into terms that explicitly protected digital ownership rights.

In the United States, the result was the DMCA which went into effect amidst protests by academic, library, and consumer groups in 1998. The provision of the DMCA most frequently invoked to protect embedded product controls and restrictions is the anti-circumvention provision. Digital content owners, publishers, and media producers saw that their attempts to prevent online piracy by embedding copy restrictions into the content itself were not enough to stop determined hackers. Building tamper-resistant media hardware and more sophisticated anti-piracy software for PCs seemed like a promising approach to content protection. In addition to making the act of copying illegal, backers of the DMCA pushed to extend the new law to cover any software, firmware, or hardware components that were designed to protect the content. The extension, which did become law, meant that consumers were henceforth prohibited from making alterations to the operation of hardware and software that they had already purchased. Any disabling of embedded controls was categorized as a form of circumvention, regardless of whether the owner intended to make copies as a result of the changes. In addition, publishing any instructions describing how to disable control functions was also defined as circumvention. These provisions dramatically increased the coverage of earlier copyright provisions and restricted the rights of owners to alter or customize their devices in ways that were previously allowable.

In the wording of the act itself, provisions against circumventing embedded controls include the following:

> No person shall circumvent a technological measure that effectively controls access to a work protected under this title.
>
> No person shall manufacture, import, offer to the public, provide, or otherwise traffic in any technology, product, service, device, component, or part thereof, that –
>
> (A) is primarily designed or produced for the purpose of circumventing a technological measure that effectively controls access to a work protected under this title;
>
> (B) has only limited commercially significant purpose or use other than to circumvent a technological measure that effectively controls access to a work protected under this title; or
>
> (C) is marketed by that person or another acting in concert with that person with that person's knowledge for use in circumventing a technological measure that effectively controls access to a work protected under this title.
>
> to "circumvent a technological measure" means to descramble a scrambled work, to decrypt an encrypted work, or otherwise to avoid, bypass, remove, deactivate, or impair a technological measure, without the authority of the copyright owner to "circumvent protection afforded by a technological measure" means avoiding, bypassing, removing, deactivating, or otherwise impairing a technological measure.
>
> *(THOMAS, 1998)*

Enterprises of all types proceeded to use the force of the DMCA to seek penalties and criminal prosecution against instances of what they regarded as attempts at illegal circumvention. Even though researchers, computer security investigations, libraries, and non-profit organizations were explicitly exempted from some of the DMCA's anti-circumvention and copy restriction provisions, copyright owners and smart product vendors have invoked the DMCA in attempts to prevent the presentation of research findings about security flaws or the methods used for certain kinds of protection, sometimes simply by threatening presenters with prosecution and more rarely by having them arrested in advance of their presentations (Electronic Frontier Foundation, 2010).

The Recording Industry Association of America (RIAA) and its members have been relentless in taking consumers to court and seeking convictions and steep fines for any copying or sharing of copyrighted digital media, filing 35,000 cases against individual consumers it claimed were illegally sharing music online, even when there was no evidence that the individual had any intention to sell copies or profit from the music sharing site they were using. In 2009 the RIAA won a suit against a woman in Minnesota who was assessed a $1.9 million fine for downloading pirated music for her personal use, the highest penalty to date in the USA (Bloomberg News, June 12, 2009). Publicity about such outsized fines may deter some consumers from downloading pirated copies of music tracks and films, but that deterrent effect may not outweigh the response of other consumers who see such prosecutions as a crusade against consumers by greedy corporate giants. The maximum fine in the United States for the illegal downloading of just one music track is $80,000, an amount that many online postings cite as being disproportional to the offense. Despite years of lawsuits against consumers by the music industry in the USA, revenue from the sale of music CDs continues to decline inexorably, dropping 14 percent between 2007 and 2008 with a total decline of 45 percent since its peak in 2000 (Sisario, 2009).

Trying to counter this decline with heavy-handed application of software and content controls may generate enough negative consumer response to force the company to back down, at least temporarily. In 2005 Sony-BMG embedded a new version of digital rights management (DRM) copy protection into its music CDs. In fall 2005, word spread around the Internet that the latest Sony-BMG music CDs contained DRM components that automatically installed themselves on consumers' PCs when a new CD was inserted. To protect its digital content, Sony seemed willing to compromise the privacy and security of its customers and their PCs without informing them. The aggressive characteristics of the Sony DRM system (XCP and MediaMax) made headlines and aroused the ire of consumer groups to the extent that Sony was forced to recall and replace the CDs and to assist consumers with removing the DRM software from their computers. As described by researchers J. Alex Halderman and Edward W. Felten:

> It was discovered that the XCP rootkit makes users' systems more vulnerable to attacks, that both CD DRM schemes install risky software components without obtaining informed consent from users, that both systems covertly transmit usage information back to the vendor or the music label, and that none of the protected discs include tools for uninstalling the software. (For these reasons, both XCP and MediaMax seem to meet the consensus definition of spyware.) These and other findings outraged many users. As the story was picked up by the popular press and public pressure built, Sony-BMG agreed to recall XCP discs from stores and to issue uninstallers for both XCP and MediaMax …
>
> *(Halderman and Felten, 2006)*

Sony's attempt to control consumer behavior by installing hardware-based controls on their PCs was the type of restrictive action that Lawrence Lessig predicted would become common on the Internet after the implementation of DMCA. In *Code and Other Laws of Cyberspace*, Lessig envisioned that the control of digital content using DRM techniques, backed up by the DMCA, would create an enforcement environment that could stifle technical innovation and shut down the possibility of disruptive new products and technologies entering the marketplace. Individuals would be unable to experiment with new ideas because even taking apart a product's embedded controls to understand how they operated could make them liable for prosecution. The result would be stagnation around the business models of market leaders (Lessig, 1999).

As the powerful consumer backlash against Sony-BMG demonstrates, the Internet has enabled new forms of online consumer information sharing that allow consumers to find out about product controls and to protest against them. With the popularity of blogs and social networks consumers today are quickly aware of vendor control controversies and highly technical users who used to communicate mainly with each other now have a direct channel to publicize product flaws and embedded restrictions as soon as they are discovered. Armed with this knowledge, consumers can decide to avoid the brands that overly restrict their options. Despite the provisions of the DMCA, it is also still easy to find instructions and services on the Internet for circumventing many forms of embedded control. In this sense, the Internet has proved to be more resilient and more resistant to control than Lessig and groups such as the Electronic Frontier Foundation had feared in 1999. While the Sony-BMG apology and recall of protected CDs represents a victory for consumers, it is also a cautionary example of the willingness of vendors to push the boundaries of control into new areas.

Exercising control through the language of product agreements is a less widely noticed but equally important shift in the demarcation of the rights associated with product ownership. Most consumers are not particularly aware of the legal language in the copyright notices, terms of use, or EULAs that are associated with online downloads, use of electronic devices, and computers. Consumers may never actually read these documents, much less compare them to earlier versions or think about the impact that complex legal restrictions may have for their use of connected smart products over time. However, the language and specifics of such documents have major implications for the rights of ownership and the enforcement of policies set by product manufacturers, online content and application stores, network providers, and content owners. These agreements give a lot more power to the vendor to control the use of smart products than in previous decades.

The changes that have taken place in this realm are an important component in the legal and business foundations for smart products. The fact that many purchase agreements for physical products as well as software and digital content are now couched in terms of licensing rather than ownership is in itself an important indicator that consumer rights have diminished and that traditional assumptions regarding ownership and fair use no longer apply. These EULAs and terms of use have been a hotly debated topic among researchers, lawyers, consumer rights advocates, and trade groups.

In her article "Regulation by Contract, Regulation by Machine" Margaret Radin notes that in the realm of Digital Rights Management Systems (DRMS) the use of embedded DRM software to enforce the provisions of a user license agreement can be considered a form of "perfect enforcement." In effect, the device is programmed to behave only as designed and it single-mindedly enforces whatever restrictions the vendor has decided to impose on the behavior of the consumer. Some of these limits may be so deeply embedded in the way the product operates that the buyer only discovers them when trying a new use, or when a particular option that always worked in earlier versions of the product no longer functions. When unexpected or unwelcome limits are encountered, there is no way to work around them; the product limitations and restrictions are built in to the product itself. According to Radin:

> A DRMS is a program that limits distribution and use of a particular item of digital content (text, video, music, software). It is elaborate technological copy-protection, with limitless possible permutations. A DRMS could prevent content from

being copied, or allowed it to be copied once and sent to one recipient, but deleted from the original recipient's hard drive. It could delete the content automatically after a set time period. It could link a copy to a particular computer, so that it would not be playable anywhere else. There are many other possibilities.

It is evident that DRMS's [sic.] – if wide deployment of them does come to pass – will attempt to accomplish by machine fiat what was previously attempted by contract. A EULA that comes with the software and says that I may not copy it for any purpose can be replaced by a DRMS that simply makes all copying impossible. A EULA that says my right to use the software will be revoked if I attempt to copy it can be replaced by a DRMS that simply disables the software when copying is attempted ... The recipient has no option to infringe and then argue fair use to a court; the safety-valve for fair use is "repealed."

*(Radin, 2004: 11)*

What Radin calls the "repeal" of fair use by the proactive and rigidly enforced rules of embedded control represents a radical change from traditional product terms of use. Smart product terms of use and license agreements are also being used to restrict the first-sale doctrine, a long-standing provision of US copyright law. That doctrine protects a consumer's right to resell something that they have legally purchased, whether it is an automobile, a physical DVD, or a printed book. First sale is also the foundation for a lot of activities we take for granted, such as library lending, an individual's right to lend something he has purchased to a friend, and the option to leave valued personal items to heirs without asking for permission from the original copyright owner or consulting the terms of use of a product's sales or licensing agreement.

Vendors of smart products can use the TOU to deny buyers their first-sale rights, including lending and inheritance practices. As one example among many smart devices that have similar terms of use, the Amazon Kindle is covered by a "License Agreement and Terms of Use" that includes the following restrictions:

> Unless specifically indicated otherwise, you may not sell, rent, lease, distribute, broadcast, sublicense or otherwise assign any rights to the Digital Content or any portion of it to any third party …
>
> You may not separate any individual component of the Device Software for use on another device or computer, may not transfer it for use on another device or use it, or any portion of it, over a network and may not sell, rent, lease, lend, distribute or sublicense or otherwise assign any rights to the Software in whole or in part.
>
> *(Amazon, 2009)*

Since the Kindle Terms of Use define the Device Software and Content as an intrinsic part of the Device, and prohibit the Device licensee (who probably thinks of himself as the Kindle owner) from any activity intended to "modify, reverse engineer, decompile or disassemble the Device or the Software, whether in whole or in part," these terms in effect prohibit anyone who purchases a Kindle from loaning it to someone else, from reselling it, and even from leaving it in their will to a relative. Resale of the e-books and other content purchased from Amazon and downloaded to the Kindle is out of the question, since there is no legal way to separate them from the device itself.

The tug-of-war between consumers and manufacturers concerning what the buyer can and cannot do with a product is not new. Over the last thirty years there have been wide-ranging debates and a raft of legal actions concerning what consumers should be allowed to do with succeeding generations of media recording and playback devices such as tape recorders, VCRs, read/write CD-ROMs, MP3 music players, and DVRs. In many cases, innovation and new products have prevailed over the fears of market leaders and content owners. In the 1970s, the film industry and the Motion Picture Association of America (MPAA) made repeated attempts to prevent the sale of video recorders and other devices that they feared would be used to copy media illegally. Finally, the 1984 Supreme Court ruling in the case of Universal Studios vs. Sony established that device manufacturers were not liable if consumers used their products to violate copyright as long as the main purpose of the product was for legal, non-copyright-infringing use. Only then did the digital recorder industry take off. When the first portable MP3 player appeared on the market in the 1990s, the RIAA went to court to try to block its manufacturer, Diamond Rio, from selling the device. A 1999 ruling upheld the principle that devices that had primary purposes besides infringement were not illegal, paving the way for millions of MP3 players including Apple's iPod (Zaret, 2003).

Even though the desire to exert control over consumer products and ecosystem relationships has a long history, connected smart products represent a qualitative change in control and enforcement. From the consumer's perspective, there is little visibility into the implications of license agreements and terms of use when he or she clicks on an online summary document. Once the agreement is made, however, it may carry the force of law, based on broad interpretations of the provisions of the DMCA and the enforceability of online contracts. New controls and restrictions can be added to the product even after it has been in use for an extended period of time.

In a 2009 debate about the relevance of his book after ten years of DMCA enforcement Lessig points out that even though pockets of consumer resistance and DRM hacking continue to make headlines, the majority of Internet users, the "bovine" citizens, as he described them in *Code*, have been more and more restricted in their options for interacting with their products over the past decade. Today Lessig says: "Whether hackers can crack the iPhone or not, the vast majority of the Internet lives life as the code permits" (Lessig, 2009).

While the Internet as a whole has not been tamed, Lessig's statement resonates in the context of connected smart products. The vast majority of smart product consumers, willingly or not, do live within the limits imposed by the embedded controls and ecosystem structures created by product vendors. Those product owners who wish to break free of embedded controls face daunting barriers in the combination of legal and technical constraints that come with many of their connected devices. It would be premature, however, to agree with Lessig that consumers are content to live within the limits of these product controls. If companies fail to provide innovative services and consumer value that is commensurate with the limits that they enforce, their extended ecosystems may not be enough to keep customers captive over the long term. The pace of change and innovation outside of market-leading product ecosystems and industry sectors is accelerating the availability of potentially disruptive products and services.

CONCLUSION

Technologies for post-purchase product and consumer behavior control are becoming more sophisticated. They are embedded deeply in a product's firmware which itself can be evolved through perpetual connectivity back to the vendor. Connected smart products may require regular updates from the vendor to function as expected, or to access new features and content. However, these updates may also impose new restrictions, delete content, and remove functionality without the explicit agreement of the owner. When embedded smart controls are buttressed by legal and contractual restrictions on consumer behavior, the prospect of device enforced control becomes a reality. As long as consumers have the choice of alternative products and ecosystems, however, embedded control does not guarantee market success for the vendor.

The next four chapters will analyze industry sectors where product intelligence and embedded controls have been introduced over time with mixed success – the intelligent automobile, smart mobile devices, the consumer-facing components of the smart electrical grid, and the connected smart home.

Each chapter will consider the current state of smart product deployment and the use of embedded controls in the industry's dominant ecosystems. The chapters will discuss pitfalls encountered in introducing smart products, potential for disruptive new business models, and innovations based on the delivery of smart services.

# 4 Intelligent automobiles

The intelligent automobile has been promised and prototyped for several decades. Like many smart product concepts, mass-market adoption of smarter automobiles often seems to be stalled just down the road. Take a closer look at the vehicles coming off the production line, however, and it becomes clear that automobiles already demonstrate many smart product characteristics. Embedded systems monitor engine performance, and manage braking and vehicle stability. Features such as cruise control, automatic variable transmission, and power steering are designed to operate cars more safely and efficiently without requiring specific driver instructions. Even tire pressure sensors rely on microcontrollers that communicate with the on-board vehicle network to warn drivers about tire problems. Advanced vehicle safety systems respond to complex inputs in real time and control critical driving maneuvers. This track record of incremental increases in embedded vehicle intelligence provides a solid foundation for the current worldwide push to develop and produce even smarter automobiles.

A review of automobile accident statistics around the world provides a sobering argument that human drivers need more help from their cars in avoiding accidents and keeping the roads safe. According to the US Department of Transportation (DOT), about six million vehicle crashes take place on American roadways annually. Drivers and passengers pay a steep price: approximately 42,000 fatalities and over three million serious injuries from vehicle accidents every year in the USA. Those accidents generate an estimated $230.6 billion dollars in direct and indirect economic costs, including vehicular damage, emergency response, not to mention the less tragic but equally costly and often dangerous congestion and traffic tie-ups that accompany many road accidents (Blincoe *et al.*, 2002).

The accident toll in Europe is equally grim, with more than 1.2 million people injured in vehicle crashes every year, including 40,000 annual fatalities (Connolly, 2009: 10–13). Japanese drivers are also accident prone. Almost one million traffic accidents are reported annually, causing an estimated twelve trillion yen in economic costs and over 61,000 fatal or serious injuries (Benedict, 2007). According to the European Union's eSafety website, over 90 percent of these accidents involve human error (Minarini, 2002). A five-year project by the DOT to analyze the distribution of accident types and the cause of accidents came to the same conclusion – most crashes occur because the driver of the vehicle makes a preventable mistake (US Department of Transportation, 2008). Drivers are easily distracted, tend to get sleepy, frequently forget to check in the mirror before changing lanes, or simply misjudge the distance between them and a car ahead. In a vehicle moving at speeds of fifty to seventy miles per hour, such behavior has serious consequences. Even at lower speeds, a few seconds of driver inattention will greatly increase the chance of an accident.

Rear-end collisions were the most common form of accidents in the DOT study, comprising about 29 percent of the annual crash total. In the majority of rear-end encounters studied, the drivers applied their brakes too late, too lightly, or never touched their brakes at all. Not surprisingly, when multiple drivers going in different directions come to a crossing, the chance of a collision increases significantly. Crashes at intersections account for about 26 percent of annual accidents in the USA. The DOT calls the next most frequent type of accident, representing 20 percent of the annual crash total, a "road departure." The cars that end up straddling highway median strips, tipping by the side of sharply angled exit ramps, or wrapped around roadside trees have suffered a road departure incident. The most frequent causes of these often fatal accidents are driver inattention, fatigue, and excessive speed. Decades of incremental auto safety improvements such as air bags, seat belts, and anti-lock brakes have whittled down the percentage of fatalities per the number of drivers and per millions of vehicle miles travelled each year, but the actual number of traffic accidents, deaths, and injuries has remained fairly constant. In the United States, for example, the Department of Transportation recorded 42,708 deaths due to motor vehicle crashes in 2006, compared to 42,065 deaths ten years earlier in 1996 (FARS Encyclopedia, 2009).

Faced with such data, it is hard to argue that fallible and easily distracted drivers are doing a good enough job making life-and-death decisions behind the wheel. The automotive industry, with encouragement, R&D funding, and sometimes mandates for new safety components from the public sector, has embraced the development of safer, more intelligent vehicles. A new generation of embedded intelligence components has endowed passenger and commercial vehicles with a broad array of smart safety and performance optimization features. Advanced safety systems are designed to scan the roadways for potentially dangerous situations and respond in a variety of ways depending on preprogrammed system parameters and manufacturer design decisions about the balance of control between the driver and the vehicle itself. These smarter driver assistance components are expected to reduce the number of accidents and fatalities by 40 percent in Europe between 2010 and 2020 (Safety Monitor, 2009) and by about 50 percent in Japan by 2015 (ITS World, 2001).

Intelligent auto safety components are referred to in the automotive industry as either ADAS (Advanced Driver Assistance Systems) or DSSS (Driving Safety Support Systems). Commercially available ADAS solutions include lane departure warning systems, vision enhancement systems for night driving, fog or other adverse weather situations, warning of other vehicles or objects detected in the vehicle's blind spots, adaptive cruise control systems that track the speed of vehicles in front, and even systems that monitor the driver's own alertness. These modules continually monitor the car's speed, road position within a lane, position relative to other objects in front, alongside, and in the rear, as well as the driver's pressure on the accelerator and brakes. Algorithms in each module's firmware calculate how all these factors combine and, for example, work to establish the safest distance from objects in front of the vehicle or determine optimal braking pressure to avoid a collision.

ADAS components are getting more intelligent and more accessible every year. Even though fewer than 5 percent of the vehicles on the road today are equipped with ADAS solutions, many auto makers offer these systems as options on new model cars (Williams *et al.*, 2008). However, the dramatic improvements in driving safety and efficiency envisioned by government agencies in Japan, Europe, and the USA will not be achieved simply by adding smart safety components to a higher number of individual vehicles. To achieve optimal safety results, cars must be able to communicate with other vehicles, to pass along data about traffic and road conditions, and about the road itself. Such communication will require upgrades to highway infrastructure and to older vehicles as well as implementation of networks to manage the exchange of critical data at high speeds. The smartest driving safety and efficiency solution may be the autonomous car – a vehicle so intelligent that its human owner could simply tell it the destination, answer a few questions about the desired route and waypoints, and relax while the car took over all the driving operations. The self-driving vehicle may sound like a distant science fiction scenario but it is an important feature of government-funded programs in Japan and the USA with target dates for implementation as soon as 2015 in Japan.

This chapter analyzes the features of ADAS options that are commercially available today, with particular attention to the control issues that ADAS implementation creates for auto makers and car owners. It then compares the ecosystem strategies of Japan and the USA for developing intelligent, connected roadway infrastructure and achieving the long-term goal of creating commercially viable autonomous vehicles. The chapter's final section discusses whether new entrants such as electric vehicle makers will disrupt the automotive industry by transforming current business models and ecosystem strategies with radically different approaches to partnerships, services, and consumer control.

## TODAY'S INTELLIGENT AUTOS: LIMITS OF ADVANCED SAFETY OPTIONS

ADAS solutions are good for business in the traditional automotive industry value chain. Smart safety technology was one of the few bright spots in the pervasive slump in automotive sales during 2008 and 2009. Auto buyers rate ADAS options positively and seem willing to pay extra for selected safety features, helping to stimulate overall market demand. The production of smarter vehicles over the past five years has resulted in high growth rates for the automotive semiconductor sector and electronic safety component manufacturers. Since ADAS solutions are only available on a small fraction of cars, there is still a lot of room to grow. Analysts predict that ADAS components will continue to enjoy a year-over-year growth rate that is almost three times the increase in overall vehicle production (Inouye, 2008).

As discussed earlier, ADAS research and development projects receive widespread government support with the expectation that smarter cars will reduce accidents as well as mitigate traffic congestion and vehicle emissions issues. ADAS implementation fits well with the auto makers' predominantly Hegemon ecosystem models. Development and rollout of safety modules are incremental processes that are controlled by the auto maker and its partners. Even though the core technologies used for embedded safety components are similar, each auto maker is able to develop brand-specific ADAS modules using proprietary, patentable methods of internal vehicle communication and data analysis together with distinct user interfaces and decision options.

For example, modules that provide lane departure monitoring and blind spot alert systems typically use some combination of video cameras mounted inside the car, on exterior bumpers or mirrors, or underneath the car. These cameras scan the road for standard lane demarcation lines and monitor the vehicle speed, steering, and turn signals; they may also use radar or other optical scanning to detect oncoming vehicles and measure the distance between them. All these elements work together to determine when and how to alert the driver to take action.

The proprietary elements from each car maker are reflected in when and how the information from these systems is communicated to the driver as advice. If the camera records that the vehicle is crossing lane markings or getting too close to the edge of the roadway, this data can be processed in conjunction with output from the car's other systems such as the turn signal, the brakes, the tires, and steering systems to calculate whether the driver is simply pulling off the highway for an exit or rest stop, is deliberately overtaking a slower vehicle, is mistakenly drifting into another lane, or has lost control of the vehicle (as indicated, for example, when the driver's hands are no longer in contact with the steering wheel). The way the system conveys information to the driver can be highly customized. If the driver has not activated the turn signal before moving across lane markers, the system may be programmed to produce some kind of escalating alert, often a sound like tires rumbling on a rough surface or a vibration of the steering wheel. Safety alerts frequently use this type of haptic feedback because the driver can process it quickly and it doesn't require him or her to take his/her eyes off the road. The vibration alert may prompt the driver to pull back into the lane, or if the lane departure is deliberate, it can serve as a reminder to use the turn signal. If the driver doesn't respond and the car keeps drifting across lanes, the system may take control of the car at some point to steer it back into the lane, or it may just make the alarm louder. In a different brand of car, the alert may be delivered as an icon in the rear-view mirror, or in a voice reminder followed by an alarm if the driver doesn't act.

Advanced safety and driver assistance features that are available in 2009 and 2010 passenger cars include:

- night vision enhancement with radar sensors
- lane departure monitoring and alerts
- blind spot monitors and alerts
- adaptive cruise control (ACC)
- collision mitigation and avoidance systems
- parking assistance with sensors
- speed limit monitoring and alerts
- driver monitoring to detect drowsiness and inattention and provide alerts.

These features are currently clustered in the higher-priced vehicles from auto makers such as BMW, Mercedes, Nissan, and Volvo. These ADAS options are expected to spread to the mid-market models of all manufacturers within the next five years (Williams *et al.*, 2008). This mirrors the rollout pattern of previous innovations such as anti-lock brakes, automatic transmission, and sophisticated sound systems which were introduced at the high end of the market and gradually became options for all car makes and models.

ADAS systems typically provide warnings and alerts to prompt the driver to take some action relative to a potentially dangerous situation rather than taking control from the driver. Many auto makers emphasize that their intelligent vehicles provide helpful safety alerts rather than mandates and state repeatedly that all decisions are up to the driver. For example, a vehicle equipped with blind spot monitoring will alert the driver with a visual display if it detects an oncoming vehicle entering the driver's blind spot and will also provide a warning if this vehicle would make a lane change unsafe. A car with intelligent speed assist will notify the driver if the car exceeds posted speed limits or is traveling too fast for prevailing road conditions. However, it's up to the driver to process these alerts and decide to take whatever corrective action may be needed to avoid an accident. He is free to ignore the warnings, speed up, or dart into another lane at any time. This freedom may be reassuring to car owners who want to keep control of every aspect of driving. However, designing ADAS interfaces that require explicit drive initiative to activate features like collision avoidance significantly reduces the likelihood that ADAS will prevent certain types of accidents.

Consider the following scenario:

> A next-generation car is proceeding down a four-lane highway at a moderate speed, keeping in the travel lane a few hundred yards behind the vehicle in front. Suddenly, the car swerves right and then back to the left, crossing into the passing lane and picking up speed while heading for the oncoming traffic lane. Inside the car, the lane warning signal has already recorded the abrupt lane cross, and the steering system has noted that the driver is no longer holding on to the steering wheel even though body pressure on the wheel has increased. This combination, together with the unresponsiveness of the driver to the urgent in-car signals, triggers a collision avoidance and automatic steering module to take control of slowing down the car and steering it back into the original travel lane after scanning the road for collision dangers. The system brings the car to a gradual stop on the shoulder of the highway. The driver is still unresponsive, so the car's emergency communication system sends a call for roadside assistance, alerting police and an ambulance that the driver may be unconscious.

The series of activities through which this fast-thinking intelligent automobile saves its driver are within the capabilities of today's ADAS solutions. But this scenario features a next-generation car because today's safety modules are programmed to default to driver control and decision making in any crisis. ADAS systems may mitigate the consequences of a collision by priming seat belts and air bags, or applying the brakes at the last second when a collision is imminent, but they will not take decisive action to avoid an accident. In practice, today's safety assistance systems would not have been enough to save an unconscious driver and may not be forceful enough to prevent chronically distracted drivers from causing accidents.

This demarcation between delegated control and pre-emptive control isn't due to insurmountable technical limitations in the capabilities of current safety systems such as driver alertness monitoring, collision avoidance, or automatic braking. Nor is it the result of a broad consensus about the safest balance between the average driver's ability to respond to an imminent collision and the best time to revert to the autonomous action of an intelligent embedded system. When radar sensors, internal cameras, and forward speed algorithms all predict that the vehicle is going to collide with an object in a matter of seconds and the driver of the vehicle doesn't give any indication of noticing the problem even after several warnings, prioritization of passenger safety would seem to dictate that immediate application of the brakes would be an obvious and highly desirable autonomous and pre-emptive system response.

This reasoning, however, doesn't take into account potential liability issues that may result from pre-emptive auto control, including the possibility that a closely following vehicle will crash into the vehicle with rapid autonomous braking unless that trailing vehicle also has a smart automated braking system. The unresponsive driver in the lead vehicle is likely going to argue that none of his overt actions contributed to the rear-end collision.

Nor does it reflect auto maker concerns about the marketability of an automobile that asserts control over the driver at a critical moment of decision. Intelligent vehicles inevitably raise questions about how much control drivers are willing to turn over to embedded safety systems. It is not yet clear that consumers will ever accept pre-emptive control by smart automobiles, despite years of data proving that driver error is the main factor in most accidents.

Today's advanced safety systems straddle a fine line between simply alerting the driver to a potential hazard and taking control of the vehicle to avoid an accident. Many currently available advanced safety systems are deliberately designed to stop short of taking control of the car. A major barrier to empowering the embedded safety system with emergency control power is the problem of liability. If an advanced safety system such as collision avoidance with braking assist does not perform as expected, auto makers want to avoid liability for any damages that result. If the vehicle triggers a chain of other accidents on the roadway, or injures a pedestrian, or the driver of the car, the auto maker is a likely target of lawsuits. If injured parties take their claims to court, and the auto maker is found liable because an advanced safety component failed to prevent the accident, that judgment could set a precedent that creates legal exposure for all future collisions involving models with the same safety module, whether the driver was unconscious or just distracted by a text message.

Aside from liability fears, car makers do not want advanced driver safety systems to negatively impact market perceptions of their brand. Automotive brands that feature the power and performance of their luxury models have long marketed the thrill of mastering a high-performance machine. It's a challenge to convince upscale consumers that sometimes their car actually knows best. A closer look at ADAS features on selected 2009 car models illustrates the delicate balance between delegated driver control and pre-emptive vehicle control that auto makers are striving to maintain.

Safety systems available for the 2009 Mercedes-Benz E Class and S Class vehicles include an infrared-based Night View system with pedestrian detection and a Lane Keeping Assist module with forward-looking cameras that monitor the position of the car on the roadway. These subsystems are intended to determine if the driver is intentionally or unintentionally moving out of the lane. These models also include the Mercedes-Benz PRE-SAFE™ and Brake Assist PLUS™ design for autonomous emergency braking. Like many collision avoidance systems, Mercedes uses long- and short-range radar sensors to monitor the road ahead and to the side. The radar system is networked to the car's braking system and to the air bag and seat belts with a precisely timed sequence of interactions that is carried out when the radar detects a likely collision. The first reaction to an imminent collision is a series of alerts that gives the driver opportunity to take control manually. For example, when the system detects a rapid decrease in the distance to another object, it presents a warning signal to the driver to apply the brakes. At the same time, the PRE-SAFE™ network communicates with the car's air bags and seat belts and prepares them for an immediate response to a collision.

The Mercedes Brake Assist PLUS™ system automatically calculates the amount of braking pressure that will be needed to stop before hitting the object. As soon as the driver touches the brake pedal, the system will increase the force to the optimal braking level, ideally stopping the car in time to avoid a crash. If the driver doesn't respond to any of the warning signals, the Brake Assist waits until 1.6 seconds before the now-unavoidable crash takes place and then automatically applies partial braking (up to 40 percent of maximum) to slow down the car in the brief time remaining before impact. If the driver still fails to take any action by the time that the vehicle is 0.6 seconds away from the collision, then the PRE-SAFE™ system finally applies maximum braking power. That's too late for any hope of avoiding a collision, but it may mitigate the damage to the vehicle and its occupants (Philips, 2009).

BMW aims to maintain its high-powered performance image, even while adding sophisticated safety features. BMW's 2009 7 Series models scored the highest rating in the iSuppli Technology Availability Index for the fourth year in a row, winning points for a plethora of advanced driver assistance and safety options (iSuppli, 2009a). The 7 Series models offer an optional lane change warning system in which radar sensors constantly monitor up to sixty meters of the road in front of, behind, and next to the car. When other vehicles are within this monitored area, a colored triangle appears on the BMW's exterior mirror. If the driver turns on the lane change indicator, this triangle will begin flashing and a "discreet vibration of the steering wheel" will provide another warning that a lane change may not be safe. It is up to the driver to decide whether to proceed. BMW's online publicity material leaves no doubt that the driver remains in control, no matter how aggressively he decides to drive. "The BMW philosophy is not to impose decisions on the driver, but rather to inform him and allow him to make his own decisions, and does not remove any responsibility from the driver to adhere to road rules or drive safely." The BMW description of its optional Speed Limit Detection system emphasizes that this function is "for information purposes only" (BMW, 2008).

Volvo's long-standing emphasis on vehicle safety innovations and the brand's traditional appeal to more conservative and safety-minded drivers may explain Volvo's willingness to promote the benefits of more pre-emptive safety systems. The Volvo X60 Concept's Collision Warning with full auto-brake senses when a pedestrian steps into the path of the car, applying 100 percent braking power in order to avoid impact. Additionally, the feature promises to prevent collisions with another car if the speed differential between the two vehicles is less than sixteen miles per hour (Volvo, 2008).

Volvo has branded this module as the City Safety™ collision avoidance system, which is standard on the XC60 and optional on other models for 2009. Volvo's promotional material clearly states that "the system determines if an impact is likely, and calculates the amount of braking force needed to avoid a collision if the driver fails to react in time. The system is able to brake the car accordingly without intervention from the driver." Nonetheless, Volvo is as insistent as every other manufacturer that the ultimate responsibility for safe driving and accident avoidance remains with the driver: "It is important to underline that City Safety does not relieve the driver of the responsibility of maintaining a safe distance to avoid a collision. The automatic braking function does not react until it considers a collision is imminent" (Volvo, 2008).

To avoid liability issues and preserve their brand images, many auto makers have settled on a model of driver safety assistance in which the driver must proactively delegate control to the ADAS components. However, if drivers do not respond to safety alerts and take the recommended action in time to avert an accident, it is hard to understand how the ambitious national goals for reducing vehicle accidents and driving fatalities can be met. Those goals require a more far-reaching and potentially disruptive shift to giving the car pre-emptive control, or to putting the next generation of intelligent vehicles more directly in charge of the driving experience.

The next section compares autonomous driving implementation strategies in the USA and Japan, and the role of the public sector in accelerating auto maker and consumer adoption of advanced safety features up to and including pre-emptive control. It provides a detailed analysis of Japan's adoption of a Federator strategy for developing and deploying smart, connected vehicles as part of an intelligent transportation system with the long-range goal of autonomous driving to improve safety and reduce traffic congestion.

## JAPAN AS AN INTELLIGENT TRANSPORT SYSTEMS (ITS) FEDERATOR

If the safety objectives espoused by government planners in Japan, Europe, and the United States are to become reality, smart autos need to be able to share control with the driver and assume direct control as needed. If a car is limited to waiting for a response from the driver, the potential for intelligent transportation applications is limited. Vehicles that communicate with the road and with each other, using integrated sensors and microprocessors to sense and respond immediately to dangerous road conditions, are better equipped to avoid collisions. If fallible human drivers cause accidents, then the safest cars and roads may be those that do not require any human decision making. This implies that the ultimate safe driving machine is the autonomous vehicle.

Gartner Research defines an autonomous vehicle as "one that can drive itself from a starting point to a predetermined destination in 'autopilot' mode using adaptive cruise control, active steering, an anti-lock braking system and GPS navigation technology." Even though many of the smart systems needed for autonomous driving are available in some form in today's cars, the required road, sensor, and high-speed two-way communication infrastructure to support autonomous driving is still characterized by Gartner as being at an "embryonic" stage. Gartner predicts that autonomous driving will require ten or more years of infrastructure build-out before it is ready for mainstream adoption (Williams *et al.*, 2008).

One step toward more autonomous vehicles is development of cars that can communicate with the roadway as they travel, receiving detailed information about road conditions and nearby vehicles. If the car is equipped for two-way communication, it may also broadcast information about the condition of the roadway to roadside communication modules. This form of communication is dubbed Vehicle to Infrastructure (V2I) and Infrastructure to Vehicle (I2V). It requires interoperable communication modules on board the car and along the road for wide-area and short-range transmissions of data. One path to implementing V2I and I2V communication is creating roads with the required communication infrastructure and equipping all vehicles with two-way communication modules. In smart, connected roadway infrastructures, communication beacons are positioned along the road to collect data and transmit it to oncoming vehicles or to a centralized authority that redistributes this information via its broader communication network. Vehicles that are entering zones with reported traffic congestion, an accident, or an area experiencing flooding or icy conditions would receive advisory alerts warning drivers to slow down or reroute. These vehicles would continually broadcast current condition reports to keep this information accurate and up to date.

An intelligent transportation infrastructure may also be based on a dynamic mesh network of Vehicle to Vehicle communications (V2V) in which a car at one point in the road will automatically transmit information about roadway conditions to cars behind it in order to provide advance warning about unseen dangers ahead. This V2V communication can also play a role in the transition to more fully automated roadways and vehicles with autonomous driving capabilities, where a lead vehicle in a convoy of smart, connected vehicles is actively controlled by a driver while the rest of the vehicles are networked into the leader and follow intelligent speed, braking, and steering commands. I2V, V2I, and V2V approaches are becoming common in Japan, where over a decade of government programs and public–private sector partnerships have created a government-supported Federator ecosystem model for implementing intelligent vehicles and roadways with full participation by the country's automotive industry leaders.

In February 2009, Tokyo's waterfront district and adjacent freeways hosted a public demonstration of the progress that the government-supported Advanced Safety Vehicle (ASV), Intelligent Transport Systems (ITS), and SmartWay projects had made towards improving automobile safety and reducing traffic accidents and injuries in Japan. The public demonstration showcased advanced safety solutions from Japan's leading automotive industry brands as well as helping the many governmental entities involved see the overall progress made to date on Japan's ambitious long-term automotive intelligence and traffic safety programs.

Beginning in 1991, Japan's Ministry of Land, Infrastructure, Transport, and Tourism has funded a series of high-tech ITS programs with industry and academic participation. Since 1996, the ITS program has followed a comprehensive, long-range plan for upgrading transportation infrastructure, on-board vehicle safety, navigation and communication systems, and creating an advanced SmartWay system that would provide a platform for real-time road-to-vehicle and vehicle-to-vehicle information delivery. Given the ambitious scope of the original program vision and the complexity of the technology developed to support each goal, the 2009 demonstration provided impressive evidence of the effectiveness of a Federator strategy in fostering intelligent networked vehicle deployment.

Additional public tests are scheduled for other urban areas in 2010 with a goal to have the integrated ITS operational by the end of

2010 and fully deployed nationally by 2011. This timeline represents the final push for Japan to realize its goal of "creating the world's safest traffic environment," which has been defined by the Road Bureau as reducing traffic fatalities to under 5,000 in all of Japan by 2012 and continuing to reduce fatalities every year after that. In addition, Japan has announced a target date of 2015 for the availability of self-driving vehicles on its smart roadways (Arino, 2007).

Nine national policy goals are identified in the ITS comprehensive plan as fundamental building blocks of intelligent transportation. These goals are:

- advances in navigation systems
- electronic toll collection systems
- assistance for safe driving
- optimization of traffic management
- increasing efficiency in road management
- support for public transportation
- increasing efficiency in commercial vehicle operations
- support for pedestrians
- support for emergency vehicle operations.

*(ITS Strategy Committee, 2003)*

The first four goals of the ITS plan have the most obvious connection to the types of advanced driving and safety systems we are analyzing. But the interrelationship of all nine goals is relevant for understanding Japan's system-wide approach to enhancing vehicle and roadway intelligence as part of an overall upgrade of the national transportation and communications infrastructure. In combination, these goals amplify the likelihood of consumer adoption and broad market penetration of intelligent vehicles in Japan.

As would be expected of a Federator ecosystem initiative, the Japanese program has a long-range, incremental adoption strategy. For the past ten years Japan has worked to integrate all of its advanced driving safety features into an interconnected ITS infrastructure project with ambient communication among vehicles equipped with special devices. It has adopted a standard networking environment, using Dedicated Short Range Communication (DSRC), to build out networks on major highways nationwide. The Japan Highway Public Corporation operates the traffic information monitors and highway and intersection beacons that communicate data over wide-area wireless networks covering the whole country. Even though different agencies and cooperative organizations have direct responsibility for specific aspects of the ITS strategic plan implementation, all of the initiatives share a common, federated technical platform and system architecture. National government agencies have led the way in developing standards for this platform and have provided incentives for auto makers and other private sector ecosystem partners to support a central system rather than embark on parallel technology development efforts. This allows multiple private sector partners to build applications and devices that are interoperable with the ITS infrastructure.

The comprehensive vision for transport infrastructure adopted in 1996 included the decision to develop a standards-based Vehicle Information and Communications System (VICS) module to be the primary anchor for delivering real-time information to the vehicle. The VICS module can be attached to any standard GPS-based navigation device to create a real-time driving information system with short-range and wide-area communications support from ITS. The navigation device does the now-familiar work of pinpointing current location and providing driving directions. Once car owners have installed a standard, government-subsidized VICS module in their vehicle, they have full access to a variety of value-added services. The ITS system can alert drivers in real time to accidents and traffic congestion, as well as weather and overall road conditions for their selected route and destination. Since the VICS module connects to the in-vehicle navigation device as well as the roadside infrastructure, if an accident or bad weather conditions have caused unexpected congestion, the VICS will post alternative routes on the navigation system display. Newer VICS modules provide multimedia data with sound and images to make it easier for the driver to absorb all the available information.

By rerouting VICS-equipped vehicles to help reduce existing congestion and avoid creating traffic gridlock, the ITS program is also addressing a national goal of reducing vehicle emissions and carbon dioxide in the atmosphere. The Ministry of Land Infrastructure and Transport estimates that VICS deployment will eliminate up to 2.4 million tons of carbon dioxide emissions by 2010. This reduction is a significant contribution to Japan's commitment to attain Kyoto Protocol targets for scaling back its national carbon footprint (Arino, 2007).

In April 1996, the Road Bureau (part of the Ministry of Land, Infrastructure, Transport, and Tourism) made the first generation of VICS units available to consumers. These VICS modules were subsidized by the Government to encourage rapid consumer adoption. Vehicle owners were responsible for purchasing the navigation system of their choice and having it installed in their cars. Owners then paid a small set-up fee to integrate their navigation device with the Government-provided VICS. Once a VICS module was installed, the various traffic and information services that it facilitates were provided for free to the consumer. As navigation systems became more common, driver adoption of VICS increased significantly. By December 2008, over 23 million vehicles in Japan had an active VICS module; as of that date 33.9 million vehicles had navigation systems. With around sixty million total passenger vehicles in Japan as of 2007, this penetration of real-time vehicle information systems was advancing in accordance with the ITS objectives.

The Road Bureau continued to push to achieve a higher rate of vehicle penetration for these on-board units along with more integration of ITS devices onto a single platform. For example, the original Electronic Toll Collection (ETC) device was a stand-alone component that did not interact directly with the vehicle navigation system. Like VICS, the ETC was a nationally standardized system, so that cars using the ETC payment module could use the device to pay tolls on any roadway. As the rate of ETC adoption increased, traffic congestion at toll road collection points decreased dramatically, generating savings in labor and traffic enforcement costs as well as reducing carbon dioxide emissions for cars that no longer had to wait in line to pay tolls.

With ETC devices becoming increasingly common, the Road Bureau decided to develop an advanced ETC device with speech capability at a cost only slightly higher than the original. This speech-enhanced unit could be integrated with the car navigation system to create a lower-cost version of the VICS/navigation system combination, or it could be used in speech-only mode without a navigation system to provide drivers with audio traffic alerts, roadway condition information, and other safety and travel data. Because the entire platform is standards-based and available to ecosystem partners for integrating new applications, the VICS and ETC platform has been extended for use as a debit system for payments in other situations besides toll collection. Updated ETC modules could, for example, be used to pay the fees for parking garages and municipal parking lots as well as paying at gas stations and ferries. This broader range of utility and applicability attracted even more drivers to install and use the ETC modules. With the national utilization rate for ETC approaching 100 percent of Japanese vehicles as of 2010, the integrated VICS ETC solution will help to bring the ITS safety services to even more drivers.

The high adoption of the VICS module is creating network effects for the longer-term ITS objectives such as deploying vehicle-to-infrastructure systems and services. The more cars that have the real-time communication units on board, the more easily new services can be distributed and quickly adopted by drivers. This progress has paved the way for a rollout of Japan's ambitious SmartWay project with features such as real-time I2V communication as well V2I and V2V exchanges of data. SmartWay goals include transmitting signals from pedestrians in crosswalks and on roadways to the vehicles to promote pedestrian safety.

The next stage in the ITS plan is to add intelligence to the vehicle itself. This initiative is taking place in an umbrella program called the Advanced Safety Vehicle (ASV) program. Japanese automobile manufacturers have enthusiastically embraced the goal of ASV implementation, in part because each ecosystem partner has been able to focus on specific aspects of innovation that are the best match for their corporate strategies and expertise. Instead of mandating what advanced safety features must be available in every vehicle, the Federator focuses on interoperability of all new features with the broader ITS infrastructure and VICS-ETC components. Interoperability makes it possible for car owners to plug some of the new safety modules into their current vehicles.

While the Japanese ITS strategy and the SmartWay infrastructure is well planned out and supported by the relevant government agencies, it's essential for the vehicle manufacturers to do their part by creating standardized and interoperable connections between various automotive subsystems. With national standards and direct government support for building smart roadways and deploying real-time vehicle communication modules, individual auto makers can leverage a sophisticated infrastructure in developing their in-vehicle modules. At the same time, the manufacturers are free to differentiate their products in the ways they decide to implement the ASV. Auto makers can deploy proprietary technology for their branded smart auto modules as long as the modules are in compliance with the national interoperability standards. Accordingly, each manufacturer emphasizes different aspects of its contribution to the ASV program. A review of the publicity releases for the 2009 Tokyo demonstration highlights how different manufacturers take advantage of this opportunity. Press releases from the leading auto maker brands describe their contributions as follows:

Honda has participated in the Advanced Safety Vehicle (ASV) project from Phase 1 (April 1991 – March 1996), and in that time has developed and commercialized a number of advanced safety systems including Honda Intelligent Driver Support System, Collision Mitigation Brake System and Intelligent Night Vision System ... Honda will also display a Japan market Odyssey equipped with Honda's latest active safety technology, Multi-view Camera System, and a Life mini-vehicle equipped with the Honda Smart Parking Assist System at ITS-Safety 2010's outdoor exhibition venue.

*(Honda, 2009)*

Mazda committed to making more eco-friendly and safer vehicles as one of the major elements of its March 2007 long-term vision for technology development, Sustainable Zoom-Zoom. Through the development of driver awareness technologies as part of its involvement in ITS initiatives, the company is striving to build fun-to-drive vehicles with abundant safety technologies that will share a sustainable traffic infrastructure with pedestrians, bicycles, and all other road users.

*(Mazda, 2009)*

Nissan Vehicle-infrastructure cooperative systems are intended to reduce the number of traffic accidents by exchanging information between vehicles and roadside communications infrastructure, and among vehicles. As part of "ITS-Safety 2010," which aims to achieve the practical application of such systems by the end of March 2011, TMC will use seven vehicles to demonstrate road-to-vehicle and vehicle-to-vehicle information-exchange technologies.

For example, to help prevent vehicle collisions at intersections on priority roads where there are no traffic signals and without a good view of oncoming traffic, the system conveys information about vehicles on side roads via vehicle-to-infrastructure communication (using optical beacons) and informs them via the vehicle's navigation system. This helps drivers recognize potential oncoming or merging vehicles.

*(Nissan, 2009)*

To support the efforts of the auto makers and promote consumer adoption of smart car safety features, the Japanese Government has taken the lead in creating the intelligent infrastructure needed to provide real-time safety and traffic information directly to vehicles anywhere on the national highways or within urban roadway systems. The Government program has also mandated a standards-based approach for design and development of VICS and ETC devices as nationally interoperable components and promoted the adoption of these on-board communication units through a low-cost installation fee with no monthly service. As a result, these core ITS modules can be installed in any vehicle without favoring any particular navigation device manufacturer. With consistent and long-term Federator ecosystem support from the Government, Japanese auto makers have been willing to invest in intelligent vehicle development, including more decisive transfer of control from the driver to the car to avoid collisions. Japan's incremental and continuous progress in implementing ITS ecosystems provides a contrast to the more erratic results of US government investment in intelligent transportation during the past two decades.

SHIFTING US SMART TRANSPORTATION STRATEGIES

Like Japan, the United States launched ambitious intelligent transportation projects in the early 1990s, with the same stated objectives for improving automotive and highway safety and dramatically reducing traffic congestion and accidents. US government agencies provided incentives and funding for automotive industry leaders to participate in a series of R&D projects to develop and test networked cars and roads beginning in the early 1990s. Even the technology vision was similar; as in Japan the early intelligent transport projects in the USA supported a combination of electronically networked smart roadways and vehicles with two-way communication plus embedded safety modules. However, shifts in the political climate in Washington and lobbying by the automotive industry led to a shift away from networked roads and towards separate projects to test various technologies for creating smarter, safer vehicles. The predominance of Hegemon models in the US automotive industry hindered the creation of a shared technology platform for I2V communication and favored intelligent vehicle projects which could be more easily commercialized by auto makers.

The National Automated Highway System Consortium (NAHSC), a public–private group sponsored by the Federal Highway Administration, was created by the 1991 Intermodal Surface Transportation Efficiency Act to guide and coordinate the Government's objectives for smarter roadways and safer vehicles and to set up demonstrations of the results as a proof of technical feasibility of automated highway solutions. The Act funded the development of a number of prototype smart vehicles and culminated in public demonstrations of a prototype Automated Highway System (AHS) in August 1997 on a designated stretch of highway north of San Diego, California. The demonstration featured a General Motors (GM) concept car – the Buick XP2000 – that "includes a conceptual guidance system that would allow the car to travel at high speed locked on sensors buried in the road." According to program reports, the demonstration was a success (AHS, 1997).

The rationale for developing smarter transportation systems and the list of expected benefits for the AHS project mirrors the issues and goals of Japan's planning for intelligent vehicles and SmartWay development that was launched in 1996. According to the AHS 1997 demo report:

> More than 40,000 lives are lost each year on our nation's highways. The annual cost to the nation is estimated at more than $137 billion. Human error is a leading factor in 90 percent of crashes. AHS promises to boost safety by reducing and/ or eliminating the element of driver error ... Information technology – "smart" highways and vehicles – offers the best solution. Automated highways will increase the safety, convenience, and overall quality of travel on US highways. Anticipated results of automated highways include: significant

reduction of traffic accidents, reduced traffic congestion, more predictable travel times for individuals and for the delivery of goods, better use of existing roadways, conservation of energy resources, and reduced emissions.

*(AHS, 1997)*

The 1997 demonstration projects were a plateau rather than a foundation for more progress in intelligent roadway implementation. AHS struggled with limited funding in the face of competing infrastructure demands and the priority of upgrading highways across the continental USA. Follow-on legislation provided a billion dollar-plus budget to fund a broad ITS initiative that included smart vehicle safety modules through 2003, but Congress voted to end the ITS Deployment Program by 2005 and overall ITS research funding was reduced. Development of networked roads and autonomous vehicles was dropped from the scaled-back ITS program in favor of funding driver assistance systems and intelligent safety components for commercial trucking as well as passenger vehicles (US Department of Transportation, 2009a).

The Department of Defense (DoD) and the Defense Advanced Research Projects Agency (DARPA) continued to support the development of autonomous vehicles with a focus on military applications. The specific goal of fielding unmanned vehicles that can function on the battlefield and within war zones was developed in response to a congressional mandate in 2001 stating that: "It shall be a goal of the Armed Forces to achieve the fielding of unmanned, remotely controlled technology such that ... by 2015, one-third of the operational ground combat vehicles are unmanned." To help meet this mandate, DARPA has reached out to the private sector and the academic world to encourage innovation on behalf of autonomous vehicle development. One of its best-known programs is the DARPA Urban Challenge, which features periodic demonstrations of autonomous ground vehicles designed by contest teams with members from both industry and academia competing against each other to complete an autonomous driving course in the fastest time. None of the vehicles that entered the first challenge in 2004 made it through the DARPA course, but there were multiple finishers in the most recent challenge in 2007. As described by DARPA, "this event required teams to build an autonomous vehicle capable of driving in traffic, performing complex maneuvers such as merging, passing, parking and negotiating intersections. This event was truly ground-breaking as [it was] the first time autonomous vehicles have inter-acted with both manned and unmanned vehicle traffic in an urban environment" (DARPA, 2009a).

DARPA defines an autonomous vehicle eligible for entry in the Urban Challenge as "a vehicle that navigates and drives entirely on its own with no human driver and no remote control. Through the use of various sensors and positioning systems, the vehicle deter-mines all the characteristics of its environment required to enable it to carry out the task it has been assigned" (DARPA, 2009b). Most DARPA military programs contain classified information, and it may be that a secure remote control system designed specifically for the battlefield was excluded from the Urban Challenge program for security reasons. For whatever reason, the DARPA guidelines do not allow I2V or V2V connectivity or the use of an intelligent, con-nected roadway infrastructure to support autonomous driving. After all, in the battlefield, military vehicles have to be able to navigate completely on their own. The emphasis on designing autonomous vehicles for military applications requires a more transformational rethinking of today's commercial vehicles than the vehicle-to-road-way connectivity that has been adopted in Japan's ITS program.

In January 2008 the online news feeds in the USA and around the world were buzzing with a rumor that Rick Wagoner, then CEO of General Motors, was poised to announce that GM would deliver a completely driverless car to the commercial market by 2015. Although that announcement never took place, Wagoner's keynote address at the Consumer Electronics Show in January 2008 did highlight GM's pro-gress in implementing advanced driving safety systems. As evidence of GM's commitment to innovation in this area, Wagoner cited the auto maker's success in the 2007 DARPA challenge, where a highly customized Chevy Tahoe was declared the winner, saying: "This type of technology, unheard of fifteen years ago, has the potential to minimize traffic jams and, more importantly, greatly reduce highway accidents and fatalities" (Wagoner, 2008). In this statement, Wagoner exhibits a peculiar amnesia about the long history of intelligent vehicle and automated highway development and demonstrations in the United States, a history in which a GM Buick played a prominent role in the AHS 1997 demonstration project of autonomous driving on a networked highway. It is a characteristic of the cyclical trends in the US auto industry that the lessons learned from earlier projects do not seem to inform current strategies for technical innovation.

The US Department of Transportation (USDOT) reorganized its strategy for intelligent vehicle development once again in 2009, announcing a new initiative called IntelliDrive that would support the integration of in-vehicle advanced safety features with investment in building an intelligent roadway infrastructure.

> In January 2009, the US Department of Transportation (USDOT) launched "IntelliDrive^SM," rebranding what was formerly referred to in the intelligent transportation systems community as Vehicle-Infrastructure Integration. The IntelliDrive program is focused on enabling a surface transportation system in which vehicles do not crash and road operators and travelers have the information they need about travel conditions. IntelliDrive will establish an information backbone for the transportation system that will immediately support applications to enhance safety and mobility and, ultimately, enable the vision of a crashless, information-rich surface transportation system. IntelliDrive will also support applications to enhance livable communities, environmental stewardship, and traveler convenience and choices.
>
> *(US Department of Transportation, 2009b)*

Even though IntelliDrive's overall goals include support for DSRC-based vehicle-to-infrastructure communication on selected roadways, USDOT's short-term priorities for IntelliDrive emphasize small, incremental improvements in driver information based on "market-ready technologies" such as mobile phones and navigation devices with connectivity capabilities. There is no indication that the program aims to create the type of Federator technology platform for connected vehicles that has been a foundation for ITS in Japan. One result of these shifts in government policy and funding is that availability of V2I and V2V communications in the USA is now almost a decade behind developments in Japan.

However, some aspects of smart vehicles and intelligent transportation services are advancing at a faster pace. As consumers begin to consider driving electric cars that plug into charging stations for power, new entrants in the electric vehicle sector are building ecosystem models that have the potential to disrupt the automotive industry. Whether or not any of these models succeed in the market over the long run, the intense interest and flood of investment dollars for electric vehicles is likely to accelerate the pace of change and pave the way for new smart product and business models in the next generation of intelligent vehicles.

## DISRUPTIVE POTENTIAL OF ELECTRIC VEHICLE ECOSYSTEMS

All-electric plug-in cars will require more than embedded intelligence and smart internal systems. In addition to offering the advanced driver safety features that are available in today's gasoline-powered vehicles, electric cars will need systems for charging the battery and helping the driver to manage its power supply. This power management may also require new styles of driving and trip planning as drivers adjust to the differences between combustion engines and electric ones, including the distance they can travel between battery charges. Broad market adoption of electric cars will require a new ecosystem with battery providers, recharging stations, electric utility partnerships, home and office power storage units, and related smart energy management products. These new requirements are a short-term barrier to mass-market adoption of electric vehicles but they also represent an opportunity for active owner engagement and the delivery of high-value services through a smarter product eco-system that is designed for disruption and rapid growth.

Hybrid and plug-in electric cars have recently assumed a prom-inent place on the product roadmaps and in the press releases of auto-motive industry leaders. Well-known auto brands such as Toyota, Nissan, and Ford have brought out hybrid vehicles and promise that completely electric cars will be commercially available within a few years. But traditional auto makers must transform themselves before they can deliver disruptive new products and their track record for insisting on Hegemon control of innovation and services is not a promising starting point for this transformation.

As John Sviokla noted in his article "In Praise of Ecosystems," many high-value, innovative consumer services could be delivered through a connected, intelligent vehicle, if only the automotive industry leaders were willing to open up their ecosystems to more innovative partnerships. Sviokla writes: "Automobiles, with their efficient, self-sustained power plants, should be platforms for tele-communications and entertainment systems; a traffic jam on the Los Angeles freeway is a telecommunications network waiting to be born. Instead, carmakers design products to accept only proprietary parts and subassemblies designed by captive suppliers – and they forgo revenue from any number of services" (Sviokla, 2005).

Would-be Charismatic Leaders and Transformers of the next generation of electric-powered smart vehicles are moving quickly to take advantage of the opportunity that traditional auto makers have largely missed – developing vehicles that will become con-nected smart platforms for multiple revenue-generating services. As new entrants, they must disrupt the business models of incum-bent auto makers. Two contenders for causing such disruption are based in California – Tesla, founded by Elon Musk, a former PayPal executive, and Fisker Automotive, founded by Henrik Fisker, previously an automotive designer with well-regarded experience at BMW and Ford. The founders of both companies are outspoken about their desire to revolutionize traditional automotive industry models for production and to create innovative ecosystem partnerships.

The committed early adopters of sustainable energy innovations are the obvious target market of Charismatic Leaders. Despite prices in the $100,000 range, Tesla has sold out its early all-electric models to wealthy believers in the cause of energy conservation. However, to move from green trophy cars to being the primary mode of daily transportation, the next generation of electric cars has to meet some significant price, performance, and driving range challenges. Ecosystem strategies are going to be essential parts of solving those challenges.

Unlike today's hybrid cars which combine efficient but traditional gasoline engines with a battery that significantly improves mileage rates, all-electric vehicles are completely dependent on battery power to keep them running. When the battery needs recharging, the driver needs to find a place to plug in and charge up just as much as the driver of a gasoline-powered car needs to find a gas station. It is an inconvenience and a bit of an embarrassment to run out of gas but thanks to the availability of roadside emergency services, the prevalence of gas stations, and the availability of convenient gas cans to transport the amount of fuel needed to get started again, an empty gas tank seldom causes more than a short delay. Running down an electric vehicle's battery could be vastly more inconvenient unless a widely available and easily located recharging infrastructure is put into place.

Fisker addresses this issue by adding a two-liter combustion engine to his vehicles, a patented drive system called Q-drive. Unlike hybrid combustion engines, the Q-drive doesn't directly power the car; it serves to recharge the battery to extend its total driving range from 50 miles on a single charge to 300 miles on the charge plus engine-assisted recharging. Fisker has adopted a disruptive ecosystem strategy as well, relying primarily on outsourcing to provide all the components and production requirements of the vehicle. Outsourcing even encompasses engineering and operations experts, leaving contractors in charge of all but a handful of in-house vehicle design decisions. The entire production process is also outsourced and production costs are further reduced by the use of standardized prebuilt components wherever possible, including many from GM and other established car manufacturers. Fisker Automotive operates with hundreds rather than thousands of full-time staff in an attempt to minimize the ramp-up costs of the new operation. Fisker intends to pass on these savings to car buyers, keeping model costs lower than other electric car makers and scaling to global sales as quickly as possible.

While Tesla and Fisker Automotive are focused on solving the production, distribution, and battery range challenges through innovative ecosystem strategies, other companies are espousing even more disruptive models.

Better Place, founded in Israel by Shai Agassi, is aiming to transform the automotive industry by creating a new model for electric car battery charging and replacement. The Better Place mission is to be the "global provider of electric vehicle services, accelerating the transition to sustainable transportation" (Better Place, 2009). Since founding the company in 2007, Agassi has convinced an impressive roster of investors and pilot implementation partners that Better Place has created the fastest and most efficient model to implement an electric plug-in vehicle that will be commercially viable. Agassi has attracted over $200 million in venture capital (VC) investment and has signed agreements with electric utilities, municipal and commercial vehicle fleet owners, national and local governments, and enterprise sectors to test the Better Place vehicle.

The basic Better Place business model is both simple and disruptive. Electric car drivers will have the option of switching out batteries at specially designed battery service stations rather than having to wait the hour-plus time it would take to fully charge the battery at electric charging stations. When convenient, such as when the car is parked at work or overnight at home, the slower recharging option can be employed. In addition to this insight, Agassi has identified the battery itself as the center of recurring services and value for the electric car. This shift led to the idea that consumers should be able to buy or lease the battery separately from buying the car; batteries are the transformational technology platform. They need to be standardized, easily removable for quick switching when drivers don't want to wait for recharging, and available for sale or lease from multiple ecosystem partners.

The insight that batteries could be leased or bought separately from the electric vehicle is getting attention from other auto industry executives. The high cost of batteries is one of the major factors in the high price tag of today's all-electric vehicles and the lease model breaks down the cost barrier and helps to reduce consumer risk in committing to a new and still-emerging technology platform. Consumers could select a battery provider or lease a Better Place-branded battery and electric vehicle. In exchange for signing a multi-year lease for the cost of the battery and perhaps a prepayment on charges and switching options, the customer could select the car model and design of their choice and get the whole package on a leased service basis.

In addition to its transformational business model, Better Place has designed a smart auto operating system that will assist drivers in optimizing vehicle performance and make both plug-in recharging and full battery-switching as convenient and seamless as possible. Dubbed AutoOS, the Better Place system includes a smart key fob that is connected to the battery to provide a visual display of the level of charge available even before the driver starts the car. Flashing blue, for example, means a full charge. The sensors in the battery are linked to the car's Better Place navigation system so the system can locate and display the nearest plug-in charging stations and battery switching bays. Once the driver inputs his destination, AutoOS will calculate whether the car can reach it on the current charge or will need to schedule recharging or a battery switch some-where en route. If switching is anticipated, the system will message ahead to set up a reservation for the fastest possible service, targeted at five minutes or less once the ecosystem is fully established.

Almost all the required infrastructure for Better Place plug-in chargers and switching stations has yet to be built out, and the vehicles themselves are still in the early prototype and testing stage. Commercial production is scheduled to start in 2011. Nonetheless, Better Place has made significant strides in winning proponents and ecosystem partners for its transformational model, not just in Israel but in many parts of the world. The largest Japanese taxi fleet operator, a utility company in Denmark, and the state of Hawaii have signed up for pilot projects starting in early 2010 (Better Place, 2009).

CONCLUSION

After decades of Hegemon domination and incremental development of driver assistance and smart safety systems, opportunities for dis-ruption in the automotive industry seem to be emerging. Models for this disruption include Charismatic Leaders such as Tesla and Fisker that are hoping to attract early adopters of electric vehicles and turn them into fans and evangelists, and Transformers such as Better Place that are aiming to revolutionize the automotive own-ership model with alternatives based on leasing the electric battery. Not all the new entrants who are aspiring to become smart ecosys-tem leaders will survive. But it is already clear that the traditional automotive Hegemon models will not set the pace for future indus-try innovation.

The next chapter will analyze the impact of smart products and disruptive ecosystem models in mobile and wireless communi-cation, another industry with a long history of Hegemon dominance and control.

# 5  Smartphones and wireless services

> The i-mode™ ecosystem brings together skills, creativity, and resources in a vast range of fields and industries, in everything from content creation to phone manufacturing and service provision … What role does DoCoMo play in the i-mode ecosystem? We are a player, not a dominant force. If we were, for example, a carnivore that devoured all the herbivores, or a herbivore that ate all the plant life, the ecosystem balance would be lost and we too would fall into ruin.
>
> (Natsuno, 2003)

As the largest wireless carrier in Japan when the original mobile app store, i-mode, was introduced in 1999, NTT DoCoMo's description of itself as just another player in i-mode's ecosystem is a bit disingenuous. DoCoMo was undeniably the dominant force behind i-mode. It is true, however, that in creating a new mobile content and service platform, DoCoMo deliberately resisted devouring all the revenue in sight. The DoCoMo business model allowed i-mode content and application providers to keep an unprecedented 91 percent share of the application revenue generated by i-mode subscribers. In addition, DoCoMo provided direct billing to its subscribers for all i-mode content, relieving content partners of the need to handle financial transactions. These business decisions made the ecosystem attractive to very small participants as well as to partners with well-established brands. Compared to the terms offered to application developers and content partners by wireless carriers in the USA and Europe at that time, DoCoMo's wireless data strategy was structured to encourage and reward widespread partner participation and innovation.

DoCoMo's technical platform decisions also facilitated rapid ecosystem growth. The format selected for mobile i-mode content was a direct subset of HTML, the basic web content formatting standard. This decision made it easier for partners to create i-mode-compatible services. On the other hand, DoCoMo maintained the Japanese carrier model of exercising strong control over mobile handset manufacturers, dictating the design features of i-mode-compatible handsets. This combination of openness and control proved to be very effective in jump-starting subscriber use of wireless data services. It ensured that i-mode phones would have an application-friendly and easy-to-use interface. Combined with generous revenue-sharing terms for content providers, it ensured that new subscribers would find a plethora of i-mode apps and services, motivating them to use the service regularly. The i-mode platform attracted more than 30 million subscribers in its first 3 years and an impressive 50,000-plus i-mode content providers and application developers (NTT DoCoMo, 2009). DoCoMo's ecosystem strategies paid off by generating significant new data revenues; like today's iPhone users, i-mode subscribers were avid consumers of mobile applications and data.

The i-mode launch and its resounding market success in Japan had other parallels with the rapid consumer adoption of Apple's iPhone. Both i-mode and the iPhone served to remind the wireless industry that mobile subscribers care very much about the content and application features available for their handsets. Both market successes demonstrated that third-party applications have the potential to become highly profitable revenue centers as well as creating competitive differentiation and subscriber loyalty for an entire mobile ecosystem and brand.

Why did iPhone's popularity and its application-centric ecosystem strategy seem like such an innovation almost ten years after DoCoMo's very successful launch of i-mode? What happened to mobile ecosystems in the interim to make the iPhone focus on mobile applications and content-based services a revelation rather than just one more example of an obvious and market-proven strategy? Instead of adopting the i-mode platform as a global content distribution and development model as DoCoMo had envisioned, the majority of wireless carriers took a different approach during the intervening decade. Carrier revenue-sharing models, mobile content formatting decisions, and hurdles for application approval and on-phone distribution relegated smaller application developers and independent content providers to the edge of the mobile network operators' Hegemon ecosystem models. Many aspiring mobile ecosystem partners starved outside the gates of the wireless carriers' walled gardens.

DoCoMo invested in KPN, AT&T, and other wireless network operators in an attempt to export the i-mode model to the rest of the world (McMillan, 2002). But DoCoMo couldn't inspire other carriers to adopt a more open ecosystem model for applications. US wireless carriers were not ready to work with tens of thousands of content providers and very small, independent application developers. After DoCoMo's investment, AT&T announced a variation on the i-mode model, invited application developers and content providers to sign up to participate, then kept most of the respondents in limbo while it shifted and adjusted the rules for revenue sharing, application submission, distribution pricing, and development guidelines. By the time the AT&T content distribution platform was ready, many of its potential partners had moved onto other application ecosystems. And there were plenty of application distribution models to choose from, since almost every major carrier ended up developing some version of a controlled application ecosystem. Whether the application development platform was based on an open and standardized technology such as J2ME (Java Mobile Edition), SMS (Short Message Service) or MMS (Multimedia Messaging Service), or a proprietary distribution platform like Qualcomm's BREW (Binary Runtime Environment for Wireless), the carrier ecosystems were difficult, expensive, and frustrating for most developers to enter. As a result, carriers failed to achieve results comparable to the subscriber and application developer popularity that i-mode continued to enjoy in Japan.

The decision of most wireless carriers in Europe and the USA not to adopt a more open ecosystem model is especially striking in the context of relentless pressure to find new sources of data revenue and to stem the decline in average revenue per user (ARPU). It was obvious at the end of the 1990s that the lifecycle of the limited capacity and slow-speed 2G digital networks was coming to an end. 2G wireless data speeds were around 9.6 kpbs, comparable to early dial-up modems. These speeds were enough for text messaging and emails, but data-intensive applications like games with rich graphics, mobile photos, and streaming music and video were not viable for subscribers with 2G service. By the year 2000, wireless carriers around the globe were already contemplating multi-billion dollar investments to acquire spectrum licenses for faster 3G cellular networks. In countries where 3G licenses were auctioned to the highest bidders, such as the UK, spectrum costs skyrocketed. In addition to the cost of 3G wireless spectrum, carriers spent billions more on the 3G-compatible equipment and switching infrastructure needed to build out higher-speed cellular networks that could handle more data traffic, mobile devices, and wireless subscribers (Sokol, 2001).

The chances of carriers recouping these network upgrade investments depended on convincing wireless subscribers to pay for more data usage, for media-intensive applications and services, and for higher-priced mobile phones. As voice-calling plans plummeted in price, and voice services became a competitive loss-leader for signing on new subscribers, wireless carriers needed to create a profitable and high-growth business model for data-hungry wireless applications as quickly as possible. Forging partnerships with as many content and application development partners as possible would seem like an appealing strategy to meet this challenge.

The habits and business modes of the Hegemon model, however, were deeply ingrained in leading wireless carriers. AT&T, Verizon, and many European wireless giants had their roots as the original monopoly operators of national and regional landline telecommunications systems. The telecom industry had long insisted on controlling every aspect of its network, including the devices that were connected to its end points in the homes and offices of landline subscribers. Only the force of the courts and the mandated opening of national monopolies to communications competition allowed the entrance of more telecom providers and wireless carriers. Up until 1966 AT&T was fighting to uphold the limits that it had imposed on customers' use of third-party equipment, specifically that "No equipment, apparatus, circuit or device not furnished by the telephone company shall be attached to or connected with the facilities furnished by the telephone company, whether physically, by induction or otherwise" (Lasar, 2008).

The particular case at stake in 1966 involved a device called the "Carterfone" a two-way modem that users could attach to their phone lines to transmit data to similarly equipped phones. AT&T lost the Carterfone decision in 1968, opening the door to dial-up modems, fax machines, and eventually access to the Internet. The opportunity to connect innovative communications devices to a landline created burgeoning new industries and sources of increased revenue for carriers as subscribers added new phone lines. In the next four decades open networks proved themselves over and over again to be a powerful strategy for unleashing new products, services, and technical innovation as well as generating rapid revenue growth. Nonetheless, the Hegemon model remained the preferred ecosystem strategy for landline and wireless network operators.

As the next section will discuss, when it came to a choice between central control of the wireless ecosystem and an open market strategy for mobile content creation and distribution of services, wireless carriers defaulted to pre-emptive control in almost all cases. In the United States, the largest carriers kept a firm grip on handset features, activation of non-cellular wireless interfaces, mobile Internet browsing, messaging formats, mobile content, and the availability and pricing of applications. In the process, the carriers also exercised pre-emptive control over the options available to their subscribers.

## HEGEMON MODELS FOR MOBILE APPLICATIONS AND SUBSCRIBER CONTROL

Carrier insistence on ruling over all aspects of the mobile ecosystem has repeatedly hindered the interoperation and as a result the consumer adoption of new services and applications. Even when messaging and billing platforms were standardized to support cellular call roaming and intercarrier billing, SMS and MMS messaging in the USA lagged far behind other regions in the world because of slow progress on intercarrier message handling and cross-network forwarding agreements.

Camera phones and MMS services, both introduced in the early 2000s, were expected to be the first of many catalysts for increased consumer use of high-speed wireless data services (Pure Mobile, 2009). Analysts predicted that MMS would overtake the global popularity of SMS as a global messaging format by 2008, driven by the desire to snap and exchange mobile digital photos with friends and family (Business Wire, 2004). As predicted, i-mode customers and subscribers to other wireless carriers in Asia Pacific adopted camera phones and began exchanging digital pictures, music tracks, and other media using their mobile phones. But European and US carriers failed to agree on a fully interoperable means of interchange for digital pictures and multimedia messages. Nokia developed its own proprietary storage formats for the pictures produced by its brand of camera phones, and many wireless carriers tweaked aspects of the MMS standard, creating problems with interoperability when messages were exchanged between different carrier networks and handsets. In addition carriers imposed high prices for sending picture messages and added to roaming charges for sending MMS messages to subscribers in different networks or regions. Even when subscribers were willing to pay these steep charges, it was impossible to know for certain if the intended recipient would be able to open an MMS message successfully (Guthery and Cronin, 2003).

The carriers also insisted on controlling the configuration of all of the SIMs and handsets that were connected to their networks.

Each carrier imposed an arduous and expensive process of mobile device certification for every component eligible for network activation. Unlike DoCoMo which used its control over handset design to insist on full-featured, low-priced devices and promised manufacturers in return a high volume of subscribers, US carriers splintered the handset market. In addition to offering multiple low-end devices that were not optimized for application and data use, carriers insisted on having exclusive rights to offer high-end phones for a set period of time. The proliferation of mobile phones with different capabilities and carrier-specific restrictions and idiosyncrasies was a constant headache for application developers who could not count on reaching a critical mass of customers if their application only ran on a handful of devices or within a single carrier's ecosystem. It was difficult and expensive to test and certify an application because each one had to be submitted to individual carriers, each of whom had their set of preferred handsets on which the application had to run and their own criteria for application testing and acceptance.

## Control of application platforms

Hegemonic control by wireless carriers frequently resulted in widespread lack of interoperability for new services. This in turn reduced consumer interest, restricted demand for advanced applications and services, and herded developers and content providers into the carriers' walled gardens. Qualcomm established the BREW application platform and distribution system for CDMA carriers. The BREW application system required all developers to pay in advance for certification on the BREW platform. Certified developers purchased specialized BREW development tools and then had to pay again to have their applications tested by third parties on each handset model before their applications would be considered by the carriers for inclusion in their application stores. Even after going through all these steps, independent developers were seldom successful in gaining featured placement in the stores. Carriers established agreements with application aggregators who served as gatekeepers between developers and application store placement, imposing yet another layer of testing, contracting, and revenue sharing that extended development time, increased development expense, and reduced ultimate developer margins.

Mobile handsets that supported applications developed in J2ME, the mobile version of Java, came with a suite of freely available software development tools that made it much easier to create applications for these phones. However, developers still had the challenge of marketing and selling their applications to mobile subscribers. Technically, J2ME applications could be downloaded directly to compatible handsets via wireless messaging or using a cabled connection between a home computer and the phone. But carrier-to-carrier handset differences in handset configuration made it difficult to ensure in advance that applications would always install correctly and work as intended. Further complicating this business model, it was difficult to convince consumers that it was safe to download applications from non-carrier, third-party online application stores. Carriers made it abundantly clear to subscribers that if any third-party application damaged or crashed the phone operating system it was the fault of the consumer and such damages were not covered under warranty. That made downloading applications from the Internet a risk most consumers were not willing to take.

These decisions to exercise strict control and limit participation in the mobile application ecosystem were underscored by a far less generous model for revenue sharing than that offered by DoCoMo. US and European carriers retained a minimum of 20 percent and sometimes more than 50 percent of the revenues generated by applications. With the addition of commissions and fees for application certifiers, testers, and aggregators who represented the carrier but charged their own fees for service, the payback period for application development was often measured in years. Carriers like Verizon refused outright to allow any free applications or sponsored content on their BREW platforms. This eliminated opportunities for promotional- and advertising-supported content to spark consumer interest in new services. Not surprisingly, these decisions favored established brands and a relatively small group of aggregators and venture-funded content providers who could afford the high price of participation in multiple closed ecosystems.

During the period that i-mode was enjoying its first flush of success in Japan the Wireless Application Protocol (WAP) was adopted by many European and US wireless carriers as a standard platform for mobile content access. WAP was developed in the late 1990s as a protocol for mobile browser applications and displaying mobile Internet content on 2G cellphones. However, unlike i-mode, the WAP format required considerable transcoding of web-based content to display properly on mobile WAP browsers. Early versions of WAP did not support the graphical interfaces and full text browsing that was already common on web pages because most 2G phones in Europe and the USA had very limited screen display capabilities. WAP relied heavily on text displays and required users to navigate through hierarchical menus to find desired content. Even though early WAP-enabled phones were heavily marketed as the gateway to mobile Internet browsing, the user experience was far different than computer-based Internet access.

WAP did have one appealing feature from the carrier's point of view. It lent itself to control through handset configuration and the ability to lock the bookmarks on the mobile browser application to specific WAP-formatted content locations. Wireless carriers preconfigured WAP browsers to connect only to the sites of preferred content partners, often charging the content providers a fee for being featured on the handset menu. Browsers were designed to make it difficult or impossible for subscribers to reach other mobile content, a version of the walled garden ecosystem strategy. Subscribers who tried mobile browsing typically didn't enjoy this restricted sample of Internet content and refused to pay premiums for data access services based on WAP. The early hype of WAP as a mobile browser did not thrive in the context of slow network speeds, monochrome handsets with tiny screens, and the requirement for costly transcoding of web content sites to make them compatible with WAP. Companies that specialized in creating WAP-formatted content failed to generate revenues and shut their doors. Brands that had paid for preferred handset placement failed to attract many visitors and lost interest in the technology. Eventually WAP evolved into a more robust back-end technology for two-way messaging as well as interactive browsing, but its early market failure soured many carriers and business enterprises on the future of mobile Internet content as a marketing and growth strategy.

### Control of mobile handsets

Carriers further insisted on control over activating any non-cellular radio connections such as Bluetooth and Wi-Fi and more recently the GPS chipsets that support location-based applications and services. In addition to frustrating application developers, this level of control often stimulated a backlash from wireless subscribers who wanted to use alternative wireless connectivity options. When Verizon Wireless decided to limit the use of the Bluetooth wireless connection in its release of popular Motorola mobile phones in 2004 and specifically to block the wireless transfer of digital pictures, ringtones, and other mobile content from the handset to a subscriber's home computer or another handset with a Bluetooth connection, consumers fought back with a class action suit against Verizon Wireless for crippling their handsets deliberately. The details of this case, and subsequent Verizon Wireless actions, provide an example of the Hegemon mobile ecosystem in action. A summary published in 2005 highlights the case against Verizon Wireless as follows:

> Verizon Wireless customers in California are suing the
> Bedminster, N.J.-based mobile phone operator for disabling
> some of the Bluetooth capabilities in a Motorola Inc. handset.
> According to the class action lawsuit, Verizon disabled some of
> the advertised Bluetooth features in Motorolas [sic.] v710 phone.

"Verizon Wireless has enjoyed enormous financial gains by marketing and selling the popular Bluetooth v710 phone then disabling almost all of its Bluetooth capabilities, resulting in a degraded phone, which requires the customer to use other Verizon paid services in place of the Bluetooth capabilities that were supposed to be part of the phones [sic.] Bluetooth features," the lawsuit said.

The v710 handset, which Verizon released in August, allows users to use their phones with a Bluetooth headset and with compatible Bluetooth car kits; however Verizon Wireless has disabled the file-sharing capability, which allows users to transfer photos or other files via Bluetooth to their PCs, printers or other devices.

*(Solheim, 2005)*

The class action suit argued that Bluetooth file-sharing capabilities were implicitly promised to consumers when Verizon Wireless advertised that the V710 would let them use Bluetooth wireless to "connect to your PC or PDA whenever and wherever you want." Verizon Wireless justified its decision to disable the file-sharing features by claiming that developers selling content on the Verizon Wireless Get it Now application store would be disadvantaged if subscribers could use Bluetooth to get free content on their phones instead of paying to use the Get It Now service. Consumer advocates argued that moving personal photos taken with the phone's mobile camera from the subscriber's handset to a PC or a friend's phone was not a service provided by Get It Now and these actions did not infringe on anyone's copyrights. Preventing subscribers from using Bluetooth connections to freely move their own personal data from one device to another did, however, force subscribers to pay whatever data and messaging fees the carrier imposed for all such data transfers on its cellular network.

Verizon Wireless settled the Bluetooth class action suit and offered affected subscribers a $25 credit or a chance to cancel their wireless subscription without paying a cancellation penalty, as long as the subscriber was willing to "sign a Certification under Penalty of Perjury that they purchased a Motorola V710 cellular telephone because they believed it would support Bluetooth object exchange or file transfer features and either own another device that supports either the Bluetooth object exchange or file transfer features or intend to purchase such a device" (Verizon Wireless, 2008).

In effect, Verizon Wireless subscribers had to swear under penalty of criminal prosecution that they already understood the workings of Bluetooth object exchange when they bought their phone and that they already owned or would definitely buy another Bluetooth device in order to be eligible to get a $25 credit from Verizon. Not surprisingly, the subscribers who had filed the suit and those who followed it expressed disappointment at this outcome. The consumer class action suit over the V710 issue was an exception. The more typical consumer reaction was to be angry at the carrier and not bother to use either the mobile camera or the crippled Bluetooth connection for much of anything, an outcome that further limited the active use of multimedia data exchange in the USA.

Verizon Wireless and other wireless carriers did not stop disabling features of handsets that might be a threat to their control over the ecosystem, though they adjusted future marketing messages to avoid promising anything that could be interpreted as delegating control to their subscribers. The reluctance to allow free wireless data transfers that led US carriers to restrict subscriber use of Bluetooth for file transfer or access to content on other devices is an example of how deeply ingrained the Hegemon strategy is in their business thinking. It has led carriers to control the use of the GPS capabilities in smartphones, limiting consumer options to fee-based navigation and other location services that are provided by the carriers themselves.

In July 2009, James Kendrick published an open letter on his blog under the heading, "Verizon: Please Stop Disabling GPS in Smartphones on Your Network" that read in part:

Why do you release phones on your network that have sophisticated GPS hardware, yet you disable it so the customer cannot use it to its full potential? Sure, you always allow it to be used with your own navigation service, but other solutions are blocked from doing so. Even solutions that are included on the smartphones by the manufacturer, for example BlackBerry Maps by RIM, are routinely blocked by you so they cannot be used as intended. The only party hurt by this blockage is your own customer.

We can only make the assumption that you block the use of the GPS hardware by third-party providers in order to promote the use of your own subscription service. I understand your job is to derive as much revenue from subscribers as possible, but to do so by disabling functionality of the phones you sell is a big disadvantage to your customer. You are playing games with your customers, and this practice needs to stop.

(Kendrick, 2009)

Until recently carrier controls on wireless subscribers in the USA included locking subsidized handsets and refusing to allow subscribers to keep their mobile phone number when they moved to another network provider. Both of these practices were challenged by consumer advocacy groups and ended by the US Federal Communications Commission (FCC) rulings. Like the previous monopoly telecommunications practices, it took lawsuits and government intervention to force carriers to support mobile phone number portability among wireline and wireless carriers in the USA (US Federal Communications Commission, 2003).

As carriers and wireless industry associations are quick to point out, wireless ecosystem control of applications, developers, and consumer options is just one side of the story. In response to an announcement in August 2009 that the FCC intended to open an investigation into the barriers to wireless innovation in the USA, Steve Largent, president of the Cellular Telecommunications and

Internet Association (CTIA), defended the wireless carriers' track record of innovation, noting:

> Whether it be the almost 100,000 applications that are now available to consumers since the opening of the first applications store 14 months ago, or the launch in the United States of the newest smart phones, or the ability of more consumers in the US than anywhere else on the planet to access the highest speed wireless networks, or the lowest price per minute of the 26 countries tracked by Merrill Lynch, or the highest minutes of use of those same 26 countries, or the fact that we have the least concentrated wireless market on the planet, or the evolution in the way services are sold – we are excited to tell the industry's story. The wireless ecosystem – from carriers, to handset manufacturers, to network providers, to operating system providers, to application developers – is evolving before our eyes and this is not the same market that it was even three years ago. In this industry, innovation is everywhere.
>
> *(Largent, 2009)*

The indicators of progress highlighted by Largent are carefully chosen to highlight the positive aspects of the US wireless landscape. The cost of mobile voice calling has decreased significantly for post-paid US subscribers and "all you can use" data plan models have replaced the confusing and expensive cost per kilobyte models. However, factors like the average cost per minute are heavily influenced by the ratio of post-paid subscriptions which are predominant in the USA. Typically post-paid wireless subscribers opt for plans with a large number of voice-calling minutes included along with unlimited night and weekend calling; when subscribers go over the number of bundled minutes they will pay a steep surcharge for each additional minute used. Prepaid mobile usage is far more common in other markets around the world. Since every minute is metered in prepaid agreements, the average cost of minutes used will be higher, even if the actual monthly cost to the consumer is lower.

Cellular networks have been upgraded to faster speeds, but the ability to access the high-speed networks mentioned by Largent refers to coverage of large metropolitan areas and populations, not to the actual number of high-speed network subscribers. Most tellingly, the recent US increase in smartphone application adoption and application store popularity cited by Largent is directly related to the success of Apple's iPhone rather than to any carrier-based innovations.

Despite recent progress, US wireless carriers have been unwilling to open their ecosystems to handset makers and application developers over the past ten years. The majority of wireless subscribers are still subject to carrier control over the content that they download, the portability of handsets from one carrier to another, and the ways that handsets are configured to use embedded features such as GPS, Bluetooth, and Wi-Fi connectivity.

Compared to the pre-emptive controls exercised by wireless carriers over their subscribers and ecosystem partners, the Hegemon leaders of the automotive industry are open-minded in letting car owners install and use third-party smart products inside their vehicles. As a concrete example of the difference between carrier control over mobile phone content and auto maker control over smart products used in cars, imagine an alternative reality in which GM and other auto makers had banded together in 1999 to prohibit the use of third-party navigational devices in all their vehicles. This action would protect sales of expensive embedded navigational modules and services like OnStar by creating a barrier to car owners who wanted to install third-party PND smart products from TomTom, Garmin, and other manufacturers. As a justification for the prohibition, the auto makers could claim that third-party GPS signals would interfere with the operation of their embedded navigational modules and other mandated automotive modules such as the OBD II system. To further protect their position as the exclusive provider of GPS-based navigational services, the auto makers could install a device in each car that jammed the GPS signal for unauthorized GPS chips. They could then protect that jamming device with anti-circumvention software. Auto makers could invoke the DMCA to prevent car owners and mechanics from disabling this device. By adding a clause to all new automobile warranties prohibiting the installation of third-party GPS products, the auto makers would incorporate their dominance of the navigational market into their contract with the car buyer.

As it happens, auto makers never attempted to implement any such control over car owners. In fact, this type of restriction seems outrageous in the automotive sector, despite its long history of Hegemon ecosystem strategies. Personal navigation devices for in-vehicle directions are now more popular than navigation systems sold by the auto makers. Consumers can currently choose among factory-installed systems, third-party branded PNDs, and GPS-enabled smartphones with navigation applications to get directions while driving. They will be free to use the next-generation navigation device in the car of their choice, and to move that device from car to car if they wish. The GM OnStar service has survived, and a new industry sector of PND manufacturers has provided an impetus to innovation and smart services. In comparison to wireless subscribers, car owners can select and install a broad range of customized solutions for their vehicles. It is impossible to know what innovative third-party wireless applications and services might have been available by now if wireless carriers in the USA had decided to open their mobile ecosystems in 1999. However, if the success of the latest generation of smartphones is any indication, we can expect the next decade will see more rapid development of smart mobile devices and disruptive wireless services.

The next section discusses the impact of smartphones and mobile application stores on wireless ecosystem strategies and the exercise of carrier control over consumer options.

SMARTPHONES AND INNOVATION

Many eyes are focused on the latest generation of smartphones and mobile application stores as harbingers of wireless innovation as well as drivers of new revenues for carriers and app store eco-systems. Research group Wireless Expertise forecasts that by 2013, annual smartphone sales will exceed 420 million units and the total number of smartphone users will be around 1.6 billion worldwide. This rate of adoption is expected to spark an increase in the value of the global mobile application market from $4.66 billion in 2009 to around $16.6 billion by 2013 (Wireless Expertise, 2009).

The explosion of mass-market attention to the iPhone's debut and the rapid growth in the number of iPhone owners exemplifies the high growth potential of a phone maker with an advanced design and a Charismatic Leader ecosystem. The sale of an estimated 45 million iPhone and iTouch devices by the summer of 2009 was an impressive feat, but it did not attract as much attention from wire-less industry leaders as the announcement that iPhone and iTouch users had downloaded 2 billion applications as of September 2009, with estimates that the downloads from Apple's App Store were gen-erating more than $200 million in revenues every week. The procliv-ity of iPhone subscribers to embrace wireless data services and in particular their eagerness to download and use mobile applications from the App Store was qualitatively different from the response of mobile consumers to previous application ecosystems in the USA.

Apple's success in attracting developers for the iPhone plat-form, aggregating over 100,000 applications on the App Store, and convincing iPhone subscribers to download billions of these apps at a highly profitable weekly pace, effectively silenced those remain-ing skeptics who had spent much of the decade speculating that the wireless industry's traditionally closed value chain would inevitably marginalize new players with direct-to-consumer business models. The popularity of the App Store at first caught carriers and early smartphone vendors such as Research In Motion (RIM) and Nokia by surprise and quickly led to a rush to set up similar application distribution platforms to capture a slice of the billions of dollars in projected mobile application and content revenues. RIM announced the sale of its fifty millionth BlackBerry device and the creation of a

BlackBerry mobile application market in 2009 and Nokia launched Ovi, its third attempt to transform itself into a Charismatic Leader by creating an application store.

While Android-based phones remained in the background for much of 2009, Google continued to convince ecosystem partners, including carrier and device manufacturers, to adopt the Android platform, leading wireless analysts to predict rapidly increasing market share for Android. As carriers rolled out new smartphone options, subscriber adoption rates confirmed that smartphones, long identified with executives and road warriors, had become the mobile device of choice for mass-market consumers.

This upsurge in popularity and consumer adoption has led to a qualitative as well as quantitative change in data usage and mobile behavior. Consumers with smartphones are much more likely to download applications, browse the Internet and view multiple web pages than consumers with narrow-focus WAP, BREW, or J2ME phones. Summarizing the behavior of T-Mobile customers who purchased Android G1 phones, T-Mobile USA CEO Robert Dotson reported that 80 percent of G1 owners browse the Web daily and four out of five download apps at least once a week, while half of G1 customers access Wi-Fi on a daily basis (Malik, 2009). All of this engagement with applications, social networks, and other types of data communication translates into higher data revenues for wireless carriers and for all smartphone ecosystem participants.

In its first year of availability, the iPhone and other application-rich smartphones also demonstrated that smarter mobile devices are highly effective marketing and advertising platforms. By the end of 2009, the iPhone accounted for more than half of all smartphone ads displayed to US consumers (Admob Report, 2009). The ability to display mobile advertising embedded in smartphone content and entertainment applications has attracted more consumer brands to develop and offer mobile content, contributing to the rapid increase in the application options available to smartphone users.

An increase in the sheer quantity of applications, however, doesn't necessarily mean that the consumer has more control over smartphone features, operation, or even the choice of what applications can be used with the phone. These decisions are still reserved by the wireless carrier, with an added layer of control now being exercised by ecosystem brand leaders such as Apple, Nokia, and RIM. Aside from a few Google Android platform adopters, smartphone manufacturers are not opening their wireless ecosystems to shared control with third-party developers or consumers.

Apple's willingness to disrupt application and mobile content distribution models stops far short of industry-wide transformation and shared control. In keeping with Apple's Charismatic Leader strategy, iPhones are configured to exercise strong controls over application developers and consumers, extending the long-standing wireless carrier insistence on controlling certain aspects of subscriber behavior. Even though thousands of developers have embraced the iPhone platform as their best opportunity to earn revenues from mobile app development in the USA, Apple's policy of reviewing every application submitted for inclusion in the App Store has generated negative reactions among a minority of application developers. Not surprisingly, those developers whose applications are rejected are the ones most likely to complain about restrictive and arbitrary behavior by Apple. One frustrated developer writes:

> Apple's App Store is the only way to share your applications with the world, and it is lorded over by an inscrutable team of guardians devoted to maintaining control over the platform. During the last four months I've spent little time working on the application itself and almost all of that time waiting for Apple to respond.
>
> A man, a plan, an App Store. It is possible to skip the App Store if you want to give your application to a friend, but even this requires getting Apple to sign off on the transfer. The iPhone wants to see a cryptographically signed note from its

mother before firing up any binary code. The ad hoc distribution mechanism puts a strict limit on 100 copies and enforces this by requiring a copy of each iPhone's unique identification code to be bundled with the digital signature. Anyone who thought that cryptography was going to liberate the world was sadly mistaken ... iPhones won't do something unless Apple allows it.

*(Wayner, 2009)*

Apple updated its licensing terms to make its restrictions on application sharing or publication outside of the App Store crystal clear to developers and to iPhone users:

> "You will not, through use of the Apple Software, services or otherwise, create any Application or other program that would disable, hack or otherwise interfere with the Security Solution, or any security, digital signing, digital rights management, verification or authentication mechanisms implemented in or by the iPhone operating system software, iPod touch operating system software, this Apple Software, any services or other Apple software or technology, or enable others to do so," reads the revised license. "Applications developed using the Apple Software may only be distributed if selected by Apple (in its sole discretion) for distribution via the App Store or for limited distribution on Registered Devices (ad hoc distribution) as contemplated in this Agreement."

*(Ankeny, 2009a)*

The US mobile content market may finally be catching up with the success generated by DoCoMo's launch of i-mode a decade ago in Japan. However, as DoCoMo has demonstrated, a more open mobile content model is not necessarily disruptive for dominant wireless carriers. Like i-mode, the iPhone strategy is a variation on control that opens the door to more partners and more mobile content – but only under the terms that are set by the Charismatic Leader of the ecosystem. DoCoMo and Apple both had reasons to disrupt the prevailing market models for application and mobile content to gain market share, but neither of them would benefit from transforming the wireless industry sector.

It is unlikely that either the current wireless carrier Hegemon strategies or controlled Charismatic Leader ecosystems will support radical innovations in the wireless industry or in consumer options for mobile communications. Apple's interests are aligned with the Hegemon carriers against mobile IP telephony and services that move consumers away from cellular networks onto competitive and more transformational mobile communications channels. RIM's BlackBerry devices are an even better match for wireless carrier business models. Since RIM opted to partner with as many wireless carriers as possible for distribution of its devices and to focus on control of the email network and the sale of BlackBerry Enterprise Servers as its business model, its highly profitable subscriber base made the BlackBerry an appealing complement rather than a perceived threat to carriers. Even though RIM and Nokia have both opened direct-to-consumer application stores, their primary ecosystem allegiance remains with the wireless carriers.

Advanced smartphone features, appealing user interfaces, and ample application store options have stimulated pent-up US consumer demand for mobile content. As long as the ecosystem models of carriers and app store leaders remain intact, it is unlikely that today's smartphones will change the dynamics of control. There are, however, signs of a more serious disruptive opportunity for new entrants who combine VoIP (Voice over Internet Protocol) and Internet-based unified communications with smart mobile services.

Perhaps the biggest driver of transformation in the smartphone platform is the increased availability of Wi-Fi-enabled handsets around the world. Current industry forecasts predict that over 300 million dual-mode cellular/Wi-Fi handsets will be shipped in 2011 and that the number of dual-mode handsets in use has doubled between 2008 and 2010 (Ramsay, 2009). Smart mobile devices with direct Wi-Fi-enabled connections to the Internet offer consumers a different path for accessing a myriad of services, content, and voice-calling options independent of the restrictions of cellular networks. If a majority of smartphone owners start to adopt these Internet calling and mobile service options, wireless carriers will be scrambling to retain control. The next section will discuss the emergence of Internet-based calling using VoIP as an alternative and potentially disruptive platform for mobile services, analyzing Skype, Jajah, and Google Talk/Google Voice as examples of three different business and ecosystem models based on VoIP for mobile phones.

TRANSFORMING WIRELESS SERVICES

As far back as 1996, the wireless industry recognized the disruptive threat of Internet-based calling. That was the year that the International Telecommunications Union (ITU) began formal work on developing an international standard for VoIP and the year that several VoIP Internet telephony companies began marketing themselves as an alternative to expensive long-distance phone calls. America's Carriers Telecommunication Association (ACTA), the industry association representing US telecommunications carriers, promptly petitioned the FCC that Internet telephony services from software companies should not be allowed in the USA and that any entity that was not a regulated common carrier should be banned from offering any services that substituted for traditional voice calling. The petition also argued that it would be against the public interest to offer free long-distance calling, with the following rationale:

> This petition concerns a new technology: a computer software product that enables a computer with Internet access to be used as a long distance telephone, carrying voice transmissions, at virtually no charge for the call … ACTA submits that the providers of this software are telecommunications carriers and, as such, should be subject to FCC regulation like all telecommunications carriers. ACTA also submits that the FCC has the authority to regulate the Internet.

ACTA submits that it is not in the public interest to permit long distance service to be given away, depriving those who must maintain the telecommunications infrastructure of the revenue to do so, and nor is it in the public interest for these select telecommunications carriers to operate outside the regulatory requirements applicable to all other carriers.

Further, ACTA claimed that if Internet telephony companies were allowed to do business as unregulated software and Internet companies, the result would threaten the future health of the entire US communications infrastructure. The petition stated:

The unfair competition created by the current unregulated bypass of the traditional means by which long distance services are sold could, if left unchecked, eventually create serious economic hardship on all existing participants in the long distance marketplace and the public which is served by those participants. Ignored, such unregulated operations will rapidly grow and create a far more significant and difficult to control "private" operational enclave of telecommunications providers and users. Such development will clearly be detrimental to the health of the nation's telecommunications industry and the maintenance of the nation's telecommunications infrastructure.

*(America's Carriers Telecommunication Association, 1996)*

After an extended period of comments and discussion, the FCC declined to rule on this petition, clearing the way for additional VoIP software providers to enter the market. Despite the dire predictions of the common carriers, the country's telecommunications infrastructure did not collapse due to competition from Internet calling services. For several years VoIP customers of computer-to-computer services were primarily early technology adopters who did not find it peculiar to speak out loud to their desktop or laptop computer as a means of making a phone call. The poor sound quality and latency problems of early VoIP connections were barriers to wider consumer adoption, as was the requirement that each party on a call had to download and install the same software client, register for a unique screen name, and go through a somewhat complex login and dialing procedure to initiate a call. Even after all this effort, the voice quality was uneven unless both parties had a high-speed broadband Internet connection. Under these constraints, it is not surprising that although the core VoIP technology was widely available from hundreds of small software companies around the world, computer-to-computer Internet calling did not become a significant market force until the mainstream consumer adoption of broadband Internet connections during the mid-2000s. These Internet broadband connections to the home were frequently provided by cable TV companies such as Comcast. It seemed a natural step for cable TV providers who had succeeded as national Internet broadband providers to add VoIP telephony to the bundle of services that they offered the home consumer. Customers of cable-bundled services could use traditional-looking landline handsets and didn't require any special software. The landline-based VoIP option quickly became the embodiment of the alternative, disruptive communication services that the common carriers had feared. Comcast and other VoIP providers looked more and more like common carriers themselves and they attracted millions of customers with easy installation, low prices, phone number portability, and unlimited long-distance voice-calling packages. The combination of cable and Internet access with unlimited voice-calling services created a package that telecoms were hard-pressed to match.

In comparison to the marketing clout of the cable companies, Skype's launch of its beta VoIP software for a free computer-to-computer calling application in August 2003 did not seem like much of a disruptive threat to traditional landline and wireless carriers. Internet voice services that required installing software on a desktop or laptop and registering for a separate screen name had not yet attracted mass-market consumers away from their landlines, much less their mobile phones. However, Skype was well positioned to grow rapidly along with the adoption of high-speed Internet connections to the home and the enterprise. It quickly added premium services to allow customers with a Skype account to make calls to landline and to mobile phones as well as to accept incoming calls from such phones on a dedicated Skype phone number. At the end of 2006 it had 136 million registered users and by summer 2009 its customer base had grown to about 450 million users around the world, justifying its claim to be one of the world's largest communication companies (Skype, 2009).

Early in 2004 Skype opened its technology platform to third-party developers, providing an API (Application Program Interface) and an array of developer tools, conferences, and contests characteristic of an industry Transformer interested in recruiting partners to an emerging ecosystem. It also made an early foray into moving Skype services from computers onto mobile devices, developing a Skype application for Windows PocketPC™ devices. By August 2005, Skype announced the commercial availability of over 400 third-party products based on Skype's API. By December 2006 Skype applications were available for 120 different Windows Mobile smartphones and PocketPC devices. By 2008 Skype, now a division of eBay, had established itself as a global leader in computer-based Internet telephony, and an innovator in extending its platform.

This technology breakout may severely disruptive for wireless carriers that depend on a controlled ecosystem where roaming voice services are still a major revenue generator and low-cost calling plans are used to attract subscribers for more profitable data services. Skype becomes a major threat to revenues if it captures a significant percentage of voice calling and data access from the cellular network. Mobile network operators have been fighting to block Skype from their subscriber's mobile phones across the world. Ever since Skype moved out of its PC-based VoIP model to take on mobile calling as a core strategy, it has become a perceived threat to the industry.

To forestall this threat, wireless carriers in many countries have refused to carry the Skype mobile application in carrier-provided mobile application stores. If subscribers seek out Skype mobile from third-party providers such as iSkoot and download it outside the carrier's walled garden, they may find that their carrier is blocking Skype calls from connecting through the wireless network, or they may discover that the carrier has exercised its pre-emptive control of the handset to deactivate the Skype application. If subscribers persist in trying to work around these restrictions to use Skype on their mobile phones, the carriers may take even more drastic action and terminate the subscriber's mobile access altogether, invoking the subscriber Terms of Use that prohibit the use of unauthorized applications on their network.

Even as the widespread deployment of Wi-Fi-enabled mobile phones such as the iPhone and new BlackBerry models give wireless consumers direct access to Internet content and a route to VoIP calling services such as Skype, carriers have tried their best to limit subscriber network access to the use of their monthly voice and data plans. T-Mobile Germany acknowledged in March 2009 that it had been blocking its subscribers from using Skype for over two years. In addition to blocking access via the network, a T-Mobile spokesman stated that the carrier would cancel all wireless network access for subscribers who tried to work around technical barriers to activate VoIP calling software, invoking contractual limits.

T-Mobile spokesperson Alexander von Schmettow told the German online site, *The Local,* that not only does the use of VoIP go against the terms of its contract with consumers, but that the "high level of traffic would hinder our network performance," and "if the Skype program didn't work properly, customers would make us responsible for it." In closing von Schmettow says simply, "Those who violate their contracts can expect to have them cancelled" (Morrison, 2009a).

The general counsel at Skype countered T-Mobile's statement with a position statement of his own, noting that in some countries wireless subscribers were unable to find any legal path to the use of Skype mobile calling since carriers in countries like Germany and France presented a united front against the use of Skype on their cellular networks. Skype's statement points out that in many countries wireless carriers are preventing their subscribers from using VoIP services:

> It's worth noting that even if German consumers wanted
> to change mobile providers, they could not: like Deutsche
> Telekom, every other German mobile operator contractually
> forbids consumers from using VoIP applications. (this [sic.] is
> the same in France, actually.) This is a real shame: many other
> operators around the world know very well that people want to
> use innovative Internet applications, like Skype, and that's the
> reason they pay their ISP to access the Internet in the first place.
>
> *(Miller, 2009)*

A few months after its first statement about blocking Skype, T-Mobile Germany reconsidered its position and announced that it would instead start charging subscribers a special fee for using the Skype mobile application, even though the Skype service operated over the Internet rather than on T-Mobile's cellular network. The carrier's justification echoed the ACTA 1996 petition: it would be unfair to other cellular subscribers and bad for T-Mobile to let a subset of subscribers make free calls, therefore the Skype users should pay extra.

> Germany's largest carrier is now backing down from its threat to
> bar VoiP [sic.] apps, such as Skype, by offering a VoIP surcharge
> that goes on sale this summer starting at 9.95 euros ($14) a
> month and up, depending on a customer's plan. For those who
> don't buy the add-on, T-Mob said it would continue to block the
> use of the apps.
>
> *(Morrison, 2009b)*

In the USA, Skype requested that the FCC follow up on its support of open Internet connections and its rulings requiring landline carriers to carry VoIP calling on an equal basis with a ruling that wireless carriers should not block mobile VoIP application software from handsets or cellular networks (Skype Communications, 2007). It also filed a comment with the Library of Congress regarding the DMCA Circumvention Provisions, noting that:

> Where carriers and handset manufacturers allow the use of third-party software applications such as Apple's iPhone App store (used on the AT&T network) or Google's Android (used on the T-Mobile network) the carriers and handset manufacturers reserve the right not to permit the use of software applications that it deems harmful to its business. For example, while it is possible to install adaptations of VoIP applications on some smartphones, carriers' Terms of Service typically block more robust "end-to-end" VoIP products that use a wireless broadband connection rather than a narrowband connection that uses the carriers' regular wireless voice minutes. The adapted versions of applications like Skype do not provide wireless consumers with the full range of innovative features that would be available if VoIP application developers were able to harness the full benefits of the wireless data plans that the consumers pay for.
>
> (Skype Communications, 2008)

Neither of Skype's requests has been granted, leaving it to battle on a case-by-case, network-by-network basis to bring its VoIP calling and other broadband applications to Wi-Fi-enabled smartphones.

Perhaps taking a lesson from Skype's battles, Jajah moved away from its original disruptive strategy for enabling free mobile calling and messaging for consumers to embrace profitable partnerships around new business models for wireless carriers. A year after its launch, Jajah was featured in a 2007 list of "the most disruptive VoIP services" because it seemed ready to take on the challenge of fighting with carriers for market share. That article summarized the core connection service that Jajah offered to end-users as follows: "On the calling side, the service acts like a VoIP provider without any equipment or downloads. You simply enter your phone number and the person you would like to call and JAHJAH [sic.] connects them over VoIP, charging per minute fees of 2.8c per minute in the US" (Ostrow, 2007).

However, by June 2009, after significant investments by leading Silicon Valley venture capital (VC) firms and Intel, Jajah shifted its business strategy markedly, describing itself as follows: "Leveraging a universal open telecommunications platform, Jajah's managed services allow mobile operators, landline carriers, cable companies, technology companies and other businesses to adopt its voice solutions with minimal investment and time to market" (Jajah, 2009).

Jajah repositioned itself as a white-label IP calling platform that allows carriers to monetize mobile VoIP by making Internet calling low cost rather than free. T-Mobile USA, for example, is deducting minutes from its subscribers' calling plans as if they were using the cellular network as well as charging a per minute rate for the calls that Jajah users make over the Internet. The 2009 agreement between Jajah and Microsoft for enterprise unified calling and private branch exchange (PBX) integration for IP bridging is also an indication of the transition of Jajah services into a platform that can monetize IP calling as a cost-effective solution for enterprise subscribers.

This focus on business and revenue generation has helped Jajah to become a notable survivor among IP telephony companies in a landscape littered with mobile VoIP failures. To avoid alienating the carriers, Jajah adopted a Federator platform and ecosystem strategy. Rather than rapid disruption of wireless industry leaders, Jajah courted these leaders to adopt the Jajah platform for VoIP. Its value proposition harnessed VoIP services for gradual and incremental subscriber adoption by making it easier for carriers and enterprise telephony companies to monetize Internet calling. As a Federator, Jajah allowed carriers to adopt selective aspects of Internet calling by creating carrier-compatible service and billing modules. At the end of 2009, Jajah's strategy paid off, as the company was acquired by Telefónica Europe for over $200 million.

Most of the other companies offering disruptive VoIP services have shut their doors for lack of revenue. However, one of those disruptive VoIP companies, Grand Central, was acquired by Google and emerged in the summer of 2009 as a core component of Google Voice, an Internet-based unified communications service with a VoIP calling application for mobile phones. Perhaps the biggest boost for Jajah's incremental Federator solution is the fear of Google Voice for Mobile. With its own mobile operating system, global brand and loyal customers, Google could develop into a truly disruptive network telephony and unified messaging option with the potential to radically transform wireless services. A Jajah-commissioned report cites Skype and Google as new entrant threats to wireless market leaders, saying:

> Subscribers communicate differently now – voice is no longer the sole or even the central mode used. Telephony must now compete with other modes, many of which are not first nature to [network] service providers. Related to this is the growing threat posed by competitors from outside the telecom sector, such as Google and Skype, who are not network-centric and lack the costly overheads of conventional telecoms ... Subscriber behaviors are changing, and their loyalties are becoming based more on the devices and applications than the service providers and their networks. This threat has become much more real since the launch of the 3G iPhone in Q3 2008. Incumbent carriers may have most of the subscribers, but they are largely missing out on where the new revenue opportunities are being created. These are coming from the new services and business models based on the Internet and IP communications – and not telephony.
>
> *(J. Arnold & Associates, 2009)*

Google Voice for Mobile is an integration of IP-based Google Talk services and the platform of the original Grand Central VoIP service. The Grand Central offering to its subscribers was to create a new central phone number and manage all other communication points through this new number. Google Voice is an advance over the basic Skype mobile functionality because of its integration with the larger suite of Google applications such as Internet-based Google Talk, IM (instant messaging), Chat, and Video Chat, and Gmail services and innovative features such as the conversion of voicemail into text files for easy review and deletion. The summer 2009 beta launch of Google Voice for BlackBerry and Android phones received positive reviews from early users, with a reviewer from the *Financial Times* noting the advantages of Google Voice on his BlackBerry and commenting: "It looks then as if Google is moving deep into telecom territory and becoming a disruptive influence in a major industry once again. Today's new mobile versions of Google Voice are the shape and interfaces of things to come" (Nuttall, 2009).

In an interview with Nuttall, Jajah's CEO Trevor Healy was happy to underscore the threatening aspect of Google Voice and proclaim the Jajah platform solution as the VoIP alternative most friendly to wireless carrier business models, saying: "Google Voice has really shaken things up. For the first time, Google has become a serious option. It works, it's good, it's really causing heartburn in the telecommunications world and operators are looking at companies like us to see how we can help" (Nuttall, 2009).

Wireless carriers and application store ecosystem companies are following a familiar script in blocking Google Voice applications from handsets and cellular networks. As of May 2009 Microsoft preemptively banned all VoIP mobile applications from its Windows Mobile Marketplace and Google's submission of its Google Voice application for the iPhone was rejected by the Apple App Store in summer 2009, at least according to Google, which petitioned the FCC to review potential anti-competitive practices by Apple and other mobile application stores. Apple responded that the Google application wasn't really rejected, just under prolonged consideration. The FCC announced in August 2009 that it would launch an investigation of wireless competitive practices in the USA, including the exclusivity agreements between handset marketers and mobile network operators such as the iPhone–AT&T deal as well as the openness of mobile application stores to third-party applications. By October 2009, perhaps prompted by the FCC interest, the Google Voice application and a VoIP application by Vonage had both been approved for sale on the iPhone App Store.

The thrust and counter-thrust of opposing players in mobile VoIP application rollouts seems likely to continue. However, the alarm expressed by landline common carriers in 1996 about the long-term implications of Internet telephony, as well as the outcome of that original petition, may well be echoed in the statements of wireless carriers today. Disruption of landline revenues and long-distance service models came from the cable companies rather than from VoIP Internet software companies, and when it came it was even deeper and more transformational than anticipated. Even the executives of the traditional common carriers are admitting out loud that their landline businesses are doomed. Ivan Seidenberg, the CEO of Verizon Communications, was quoted at a Goldman Sachs investor conference in September 2009 as saying that "Verizon was simply no longer concerned with telephones that are connected with wires" (Hansell, 2009).

If mobile VoIP services create a similar level of disruption for wireless voice services, the industry leaders who are still focused on blocking specific applications like Skype or Google Voice would do better to begin planning for a VoIP alternative that they can incorporate into their own ecosystem strategies. Amid growing consensus that the adoption of VoIP on mobile phones is inevitable in the next five years, industry analysts are predicting that by 2013 mobile VoIP applications will generate annual revenues of $32.2 billion, with over 278 million users worldwide (Dickson, 2009). Jajah adopted the role of the incremental VoIP Federator in facilitating this growth with a platform that allows carriers to preserve some revenue and keep their customers in the fold. But the Hegemon model is still strongly entrenched in the wireless industry and it has been difficult for industry leaders to meet potential Transformers halfway.

Google is well positioned to take on the role of Transformer, building on its investment in Internet-based telephony and unified communications including mobile VoIP services. Together with the Google Android operating system for mobile hardware, its continuing interest in wireless spectrum, and its ability to monetize all types of communications with advertising services, Google has the global reach and capacity to take on wireless industry leaders over the long term. If Google or another Transformer succeeds, smarter wireless services may develop completely outside of the cellular telephony model we know today. Instead of just a smartphone, consumers may have the option of selecting the unified wireless service platform of their choice and have it available on any number of mobile and fixed devices, including their smart, connected automobile, music player, or e-book reader.

Transformational changes in wireless services may be another decade in the making. There are strong reasons for mobile telephony users to maintain a subscription to traditional cellular wireless carriers and not to rely completely on IP calling, just as millions of home phones are still based on fixed-line networks. But the momentum for wireless transformation is building. If the wireless carriers do not open up their Hegemon modes, new industry entrants have a good chance of transforming mobile communications from a phone-centric and voice-centric service to just another smart wireless service among many options on a wireless and always-connected smart product. This will change the nature of wireless ecosystems and increase the number of alternatives that consumers have in integrating all types of wireless services.

Mobile content with advertising and applications with embedded functionality are becoming a major part of consumer mobile devices. Add this to the billions of text messages, IM packets, and emails that are already sent via mobile. Voice calling is becoming just one of the many applications and as the Jajah report noted, it is not even the most important for many young consumers. Unified mobile messaging, voice, and even two-way video communications will be integrated into any number of smart mobile devices in the next five years.

## CONCLUSION

Wireless carriers are understandably concerned about the elements of the mobile value chain that are most directly related to network access and usage. However, application developers and the rising generation of mobile consumers do not see the primary value of the communications services on their smartphones coming from a specific type of cellular network infrastructure. Today's wireless subscribers assume that the basic functionality of the mobile phone will work without them having to think about it. When it doesn't work they will be irritated at their wireless carrier and consider switching. Consumer value perception has shifted radically during the past five years from selecting mobile phones primarily for voice quality and reliability to evaluating the whole package of mobile value, including applications and messaging services, Wi-Fi and Bluetooth, GPS navigation capabilities, and the quality of an embedded digital camera and music player.

Inevitably voice telephony is becoming just one of the many smart services and applications for wireless devices. In this new competitive environment, the ownership of licensed spectrum will become a less powerful foundation for control of an entire ecosystem. Companies will compete on offering the most attractive combination of connected smart devices, content, and services, including voice and two-way video options.

# 6    Energy: imbalance of power

Attached to almost every house, apartment building, and office complex in the United States is a familiar and rather homely device. Inside its glass cover, a counter ticks off the kilowatts as the meter measures and records the amount of energy consumed by kitchen appliances, air conditioners, computers, light bulbs, and other electrically-powered devices inside the premises. For more than a hundred years the basic electric meter design focused on this mission of accurate measurement, providing utility companies with a reliable record of electricity usage that could be turned into a monthly bill for each customer.

When meters finally entered the digital age during the 1990s, the priority for utilities was automating the time-consuming and costly process of meter reading rather than giving their customers more information about their power usage. The possibility that the humble electric meter could cooperate with the average residential consumer in managing peak energy demand cycles and avoiding brown-outs was, at best, a far-fetched vision for the distant future. To the extent that records of monthly electric power consumption at the residential level were used to forecast future demand and power-generation needs, the utility's strategy was straightforward. The electric utilities simply planned for continued year-over-year increases in energy consumption. The historical data for electric power demand in the United States certainly supported the assumption that consumption and demand would continue to increase; according to the Department of Energy (DoE), the demand for power rose almost 30 percent between 1988 and 1998 (US Department of Energy, 2008a).

The discouraging history of prior efforts to motivate US consumers to monitor and conserve their energy consumption seems to justify a skeptical view of consumer-based reduction in demand for electricity. Attempts to engage widespread consumer interest in the details of electricity usage have attracted only a tiny percentage of residential power customers in the past several decades. Until very recently, home automation kits and software programs that analyzed the efficiency and cost of appliance operation sold only to a limited market of dedicated conservationists. During the 1980s and 1990s, utilities in the USA that offered their residential customers the option to be billed based on variable Time of Use (TOU) rates had average participation rates of less than 1 percent (Brockway, 2008: 69). And even that minimal amount of consumer interest tended to drop off considerably over time. An Electric Power Research Institute (EPRI) study noted that one large utility offering TOU options signed up over 26,000 consumers when the program started in the 1980s, but reported that by 2004 only 11 customers were still participating (Abbott, 2005: 6–8).

Suddenly, however, the electric meter has a starring role in a global movement to transform the traditional system of power generation, transmission, and consumption. As new networking and computing technologies enhance the capabilities of national and regional power grids to turn them into interactive, smart grids, a new generation of smart meters are replacing today's models. Whether or not consumers want them, new meters will soon be attached to their homes. What remains to be seen is whether smart, connected meters will motivate consumers to change their power consumption habits. An even more important question is whether electric utilities will leverage the capabilities of smart meters to encourage active consumer participation in home energy management. Billions of dollars are riding on the outcome.

The cost of making the switch to more intelligent residential meters is considerable. Southern California Edison's project to install 5.3 million smart electric meters and meter communication equipment for its residential and commercial customers between 2009 and 2012 will cost an estimated $1.63 billion. The Obama administration announced that the Government's 2009 economic stimulus package will provide $5.3 billion in new funding for smart energy initiatives, including a target of upgrading over forty million meters in the United States. Large though it is that budget estimate represents just a small fraction of the projected cost of all the interconnecting elements of the electrical system required to create an end-to-end smart grid infrastructure for the entire United States. The Brattle Group estimates that the total cost of smart grid build-out between 2010 and 2030 will be around $1.5 trillion (Fox-Penner *et al.*, 2008).

From a public sector policy perspective, there are multiple motivations for upgrading the energy grid in the United States just as in all industrialized countries. One of the most immediate imperatives is that power-generation supply is forecast to fall behind the demand for power consumption sometime in the next decade (Energy Information Administration, 2008). Equally compelling, the generation of electric power by traditional means such as coal-fired power plants creates harmful pollutants and contributes substantially to the carbon emissions that are thought to be a factor in global climate change. Reduction in the carbon emissions from automobiles and other vehicles is a popular target in campaigns to reduce pollution, but the role of power generation is less widely publicized. In fact, the entire transportation sector in the United States emits just 20 percent of the country's total annual carbon dioxide emissions, while the generation and distribution of electrical power accounts for about 40 percent of the total (US Department of Energy, 2008a: 20).

According to a 2007 report from the US Department of Energy:

> A substantial gap exists between existing and desired
> amounts of DER (Distributed Energy Resources), particularly
> in the renewable category. Most of our electricity comes
> from centralized plants. Of the distributed generation we
> do have, most is "dirty" (from internal combustion engines)

and disconnected from the grid. The US is dominated by big, centralized generating facilities. Large generators (coal, nuclear, and hydro) made up over seventy-five percent of net generation in 2004.

*(US Department of Energy, 2007)*

Despite strong government interest in switching from centralized and "dirty" distributed power generation to renewable alternative energy sources such as wind, solar, and geothermal, the current power grid is not designed to accommodate alternative sources of energy, particularly energy that is generated at the level of individual businesses or residences. In addition to today's power-generation capacity and emissions problems, there are significant inefficiencies all along the current network of distribution substations and lines. An estimated 20 percent of all the power generated centrally is lost as electricity is transmitted over the US grid to distant substations, in effect wasting that energy and the resources used to create it. The current grid is also subject to outages and brown-outs, particularly at times of peak demand. Because of its design, a single outage can cascade across a broad area before it can be repaired. Nor can the energy that is generated be stored for future use. In the face of such limitations, the plan for a systematic transformation of the grid infrastructure, rather than spending $100 billion over the next ten years on a patchwork of fixes, has received widespread public and private sector support (Fox-Penner *et al.*, 2008).

The Modern Grid Initiative Report from the Department of Energy spells out in non-technical terms some of the characteristics that define a national smart grid:

**Self Heals** A self-healing modern grid detects and responds to routine problems and quickly recovers if they occur, minimizing downtime and financial loss.

**Motivates and Includes the Consumer** With a modern grid, commercial, industrial, and residential energy consumers will have visibility into prices and the ability to choose a program and a price that best suits their needs.

**Resists Attack** Security is built in from the ground up in a modern grid.

**Provides Power Quality for 21st Century Needs** A modern grid provides electricity free of sags, spikes, disturbances, and interruptions. It is suitable for the data centers, computers, electronics, and robotic manufacturing that will power our future economy.

**Accommodates All Generation and Storage Options** A modern grid allows plug-and-play interconnection to practically any source of power, including renewable energy sources and storage.

**Enables Markets** A modern grid supports consistent operation from coast to coast while allowing innovation locally and regionally.

**Optimizes Assets and Operates Efficiently** A modern grid allows us to put more power through existing systems, build less new infrastructure and spend less to operate and maintain the grid.

*(US Department of Energy, 2008b)*

Although the DoE definition of a smart grid includes giving consumers more visibility into electricity prices and implies that there will be more choices available, it does not say how much control residential customers will have over their electric bills as a result of smart meter installation. In fact, the DoE carefully avoids any promise of cost savings for consumers in its discussion of expected smart grid benefits, limiting itself to the more general hope that the smart grid will provide a variety of options that will meet the needs of residential customers. A meaningful level of consumer involvement in shaping those options is far from guaranteed. The energy sector today is dominated by traditional utility companies presiding over deeply entrenched Hegemon ecosystems. These utilities exercise strict control over how customers are billed and how customers can connect to – and interact with – the power grid. The impact of that utility company dominance can be seen in the priorities to date for smart grid deployment and smart meter implementation.

Most of the benefits touted for the smart grid will actually accrue to the electric utilities themselves as well as to their large commercial customers. The utilities will benefit from a substantial reduction in the need to invest in new power-generation infrastructure due to greater efficiency and lower loss rates for the more efficient distribution enabled by a smart grid. It is also in the interests of the utilities that connected smart meters provide them with the ability to do load balancing across a large percentage of their customers at times of peak demand. While everyone benefits from improvements in the quality and reliability of the grid, corporate customers have the most to gain from higher-quality power with fewer surges and power lags and from a dramatic reduction in the unplanned power outages and brown-outs that disrupt productivity and today cost US companies an estimated $150 billion annually (US Department of Energy, 2009a).

The benefits that a smart grid offers to the individual consumer are neither as clear nor as readily quantified as the benefits it offers to the electric utilities. In the midst of US Government commitments to smart meter installation and industry endorsements for smart grid implementation, it is easy to lose sight of the residential electricity customer. Few consumers have actually requested the smart meters that are turning up in their homes. Nor do consumers have an option to refuse to participate in the upgrade to smart meters, even though it often comes with a new electricity rate plan that amortizes the cost of the meter and its installation as a surcharge on their monthly electric bill. This lack of direct involvement to date makes it difficult to predict whether the majority of consumers will have the motivation and opportunity to take a more active role in personal energy management through smart products and services. Consumer surveys and residential smart meter pilot projects provide only partial and sometimes contradictory indicators about long-term energy consumption outcomes. Such pilot projects are, however, a useful source of information about the operation of residential smart meters and their limitations as a source of consumer information about energy consumption.

This chapter analyzes the impact of smart grid and smart meter implementation in the USA on four dimensions. The first section discusses the capabilities and limits of residential smart meters and the push to install such meters in order to enable variable energy pricing at the residential level. The next section reviews the results of selected utility pilot programs that have tested variable electricity pricing options with consumer volunteers and the issues that prevent such pilots from being reliable predictors of future consumer behavior. The third section discusses alternative smart energy programs that encourage consumers to self-monitor their energy consumption and allow them to generate their own power from renewable sources such as solar panels. The chapter concludes with an overview of current utility ecosystem models and the barriers to smart, connected energy products and services for consumers in today's industry landscape.

## SMART METER CAPABILITIES AND VARIABLE PRICING MODELS

Smart meters are an incremental step in a long history of electrical power measurement. Remote meter reading and other advances in metering intelligence are included under the general term Advanced Metering Infrastructure (AMI). According to a study of smart metering in Europe, smart meters provide the following three basic functions: they "measure the electricity used (or generated), remotely switch the customer off and remotely control the maximum electricity consumption." These three capabilities are much more significant for their impact on utility company operations, pricing models, and control over customers than for the technical innovation that they represent. In fact, the authors of the smart metering study note that large-scale national projects in Europe confirmed as early as 2006 that there were no technical barriers to smart meter adoption and that the technology was mature (van Gerwen *et al.*, 2006: 1–9).

What then makes this new generation of electric smart meters so important to the implementation of the smart grid and to accomplishing national goals for energy efficiency? Smart meters are essential because they give utility companies a new level of real-time control over customers. As we have seen with other connected smart products, embedded intelligence combined with networked connections enables qualitative changes in service delivery, remote updates, vendor control, and equipment performance monitoring. With a real-time connection to the utility, a smart meter can automatically communicate details about power outages and other problems to the utility on behalf of the customer. Continual remote monitoring of meters and other equipment for signs of breakdowns may even allow utilities to proactively repair problems to avoid an outage. Meters can be disconnected from a central location when an account is closed. The meters also provide automatic, detailed reports on power usage, further automating the task of meter reading and improved billing accuracy. These features reduce the utility's operational costs significantly.

Another key role of smart meters in the smart grid architecture is enabling variable pricing programs and automatic load balancing. To motivate customers to cut back on energy consumption voluntarily during periods of peak demand, utilities can raise the price of power during those periods and implement the price increase in real time by communicating it to the smart meter. And if customers don't cut back enough to reduce demand, the smart meter can be programmed to pre-emptively reduce the power available to the customer premises. This level of control significantly reduces the need for utilities to build new power-generation plants to meet future peak demand situations.

Embedded programs in the meters and connected infrastructure can be updated as needed by the utility, giving them increased flexibility in offering pricing options based on current demand conditions. A marketing brochure from Echelon Corporation, a smart meter manufacturer, highlights the key selling points for utility customers such as measuring and logging energy consumption at more frequent intervals and enabling the utilities to establish a number of variable rate pricing plans.

Echelon lists the following features of its Poly-Phase residential meter:

- time-of-use metering with up to four tariffs and customer billing cycles
- four tariffs with ten possible tier switches per day; four seasons per perpetual calendar (set by Day/Month) perpetual holiday calendar for up to fifteen holidays per year; perpetual daylight savings changeover; two separate holiday day schedules per season; one weekday, one Saturday and one Sunday day schedule per season
- power quality measurements, including outage detection and duration
- load profiling capability captures up to eight values at adjustable intervals
- automatic periodic configurable self-reads, and stores up to twelve or twenty-four sets of readings (dependent on model)
- event log with circular memory buffer to store up to a hundred events
- maximum power limiting to disconnect load when configurable power threshold is exceeded
- energy-credit-based prepay functionality, including varying deductions per time-of-use scheduling, configurable emergency credit and audible low-credit alarm
- extensive tamper detection features
- data logging intervals user selected as 5, 15, 30, 60 minutes or 1 day.

*(Echelon Corporation, 2009)*

The ability to acquire usage statistics at frequent intervals and to accommodate a variety of pricing and tariff rates gives the utility company a broad range of options in establishing variable pricing models and billing plans. This fine-grain information about energy usage is not typically made available to the consumer, however. Smart meters are designed to send all data that they collect directly to the utility company and not to communicate with the residential customer. Unless the utility provides its customers with additional products such as smart thermostats that can communicate directly with data generated by the smart meter or wireless display panels that show patterns of consumption, consumers are left on their own to figure out how to manage energy usage inside their home.

Widespread installation of smart meters gives a utility company a new way to control aggregated power consumption across its residential and commercial customer base. Energy consumption studies have demonstrated that if enough customers cut back their power usage during a high demand period even relatively small reductions in electricity usage by each customer will aggregate across the region to smooth out the overall demand cycle. In the Energy Policy Act of 2005, the US Government specifically required the Federal Energy Regulatory Commission (FERC) to work toward removing "regulatory barriers to improved customer participation in demand response, peak reduction, and critical period pricing programs" as a means to increase participation by consumers in demand-response (DR) programs (Federal Energy Regulatory Commission, 2006: 1). Studies have also shown that utilities need to use a combination of pricing incentives and direct control over the maximum power available to selected customers to ensure that enough customers will act to reduce power consumption (Federal Energy Regulatory Commission, 2006: 1).

DR programs recruit customers who agree to automatic reductions in their power during critical demand periods. The US Department of Energy defines demand response as follows: "Changes in electric usage by end-use customers from their normal consumption patterns in response to changes in the price of electricity over time, or to incentive payments designed to induce lower electricity use at times of high wholesale market prices or when system reliability is jeopardized" (US Department of Energy, 2006). The utility companies can count the reserves represented by participants in an automated DR program as a part of their peak consumption power-generation capacity. So instead of having to build new power plants or contract to pay premium rates for access to off-grid power capacity the utility can balance its load in real time by automatically reducing the power consumption of participating customers.

Prior to the installation of millions of residential smart meters, it was much more efficient for utilities to manage a DR program with a small number of selected commercial customers. Standard home meters did not provide the ability for utility companies to implement demand reductions in residential power consumption, nor did they support a variable pricing structure that could charge more for the electricity used during peak demand periods. Enterprises that consumed high amounts of power on a continual basis, such as commercial customers with large clusters of buildings and equipment or energy-intensive manufacturing and production processes, were the prime candidates for DR incentives. The participation of these customers was so valuable that utilities often motivated them to sign up for what is termed an interruptible/curtailable service by providing financial incentives and free installation of commercial smart meters. Commercial DR participants are eligible for a monthly check from the utility whether or not they are actually called on to cycle down their power consumption during the month.

Installation of residential smart meters makes it realistic for utility companies to include their residential customer base in calculations about load balancing and demand management. However, utilities do not provide their residential customers with the financial incentives available to the commercial entities that are offered interruptible service payments. DR pricing programs for consumers are designed to offer a very small carrot in the form of slightly lower prices during off-peak periods combined with the formidable financial stick of critical peak prices that may be up to ten times higher than normal rates. Customers who sign up for variable pricing plans or DR billing receive advance notice from their utility company when a critical peak pricing period is about to go into effect. If they respond by cutting back on their power consumption or shifting it to non-premium periods, by either actively turning off or cycling power usage by home appliances or giving the utility company permission to do so on their behalf, the reward will be stable or possibly lower electric bills. If customers do not reduce their consumption when premium pricing is in effect, however, they will end up with higher bills.

Both the benefits and the potential penalties embedded in residential variable pricing and DR rate programs are premised on utilities receiving approval from regulators to change their flat rate price structure. Traditional flat rates for electricity are based on complex formulas that average the amounts paid by the utility to generate its own power, amounts paid to purchase power from aggregators at wholesale prices during periods of normal demand, and amounts paid to obtain additional power at premium rates during periods when demand increases beyond normal capacity. The aggregated flat rate structure used to bill residential customers also includes the costs of power transmission and distribution and a charge for the long-term recovery of – and profit on – capital investments that the utility has made.

Whenever costs go up or profits go down, utilities can propose new rate structures to their state regulators; such proposals will typically request a rate increase for the residential consumer. Rate structure proposals can also include requests for approval of variable pricing and DR pricing options. While the majority of US consumers today are billed based on the traditional, cost-plus flat rate plans, variable pricing and DR programs are gaining popularity, spurred by the provisions of The Energy Independence and Security Act of 2007 which sets milestones for state adoption of demand response. This Act also required that FERC conduct a national assessment of DR potential and report the results to Congress. The first assessment report, "A National Assessment of Demand Response Potential," published in June 2009, discusses various means by which consumers can be convinced to participate in DR programs. It is worth quoting at length as an illustration of the relative lack of consumer control over smart metering and other utility company decisions.

Much of the assessment report addresses concerns that utility companies will receive less revenue from consumers if smart meter installation and DR program implementation succeed in creating significant decreases in residential power consumption. According to the report:

> Without certain regulatory mechanisms in place, utilities generally have a disincentive to pursue programs that will reduce sales. While this problem is most pronounced with energy efficiency programs, it is also present with programs to encourage demand response. Ultimately the reduction in sales that results from demand-response programs will cause the utility to fall short of receiving the fixed revenue requirements that would otherwise be recovered in the absence of the sales reduction.
>
> *(Federal Energy Regulatory Commission, 2009: 197)*

One solution discussed in the report is "direct cost recovery" meaning that the utilities simply charge their customers enough extra every month to make back the money that consumers may save through reducing their power consumption in a DR program participation. While direct cost recovery is the most common solution approved by state regulators, the report notes that this strategy is not as effective as "fixed cost recovery" which allows utilities to increase rates over time to make up for any shortfall in cost recovery plus profit margin associated with decreased customer consumption of energy.

The report goes on to note that customers who anticipate receiving bill increases might not participate voluntarily in a DR program, cutting down on the potential reduction in consumption:

> Additionally, dynamic rates by definition will result in some customers experiencing bill increases due to their

> peakier-than-average consumption patterns, and these customer
> may not opt-in to such a rate if it is only offered on a voluntary
> basis.
>
> *(Federal Energy Regulatory Commission, 2009: 200)*

To address the issue of consumer reluctance to participate voluntarily, the report presents two scenarios for consumer participation as follows:

> The Achievable Participation scenario assumes universal
> adoption of default tariffs that impose dynamic pricing on
> customers unless they expressly choose not to participate in the
> program. The Full Participation scenario assumes mandatory
> participation in dynamic pricing programs by all customers. To
> achieve the estimated demand response potential under either
> scenario, it would be necessary for retail regulators to modify
> existing electric utility rates and rate structures to implement
> dynamic pricing on a default or mandatory basis. Such rates
> would need to be designed to ensure that dynamic prices provide
> for adequate recovery of investments, while also offering time-
> varying electricity prices to customers.
>
> *(Federal Energy Regulatory Commission, 2009: 73)*

The FERC report puts the implementation of smart meters into a macroeconomic and social context. Federal energy priorities include reducing national demand for energy in the cause of energy independence and security, saving on the cost and carbon impact of building new "dirty" power plants, and preserving the current business model of utility companies. To the extent that smart meters and DR programs disrupt current business models, their impact will be mitigated to make up to the utility companies for any lost revenues. To the extent that consumers don't voluntarily cut back on electricity consumption in a DR program, they will pay a steep financial penalty in higher electricity bills. This economic and political outlook is a far cry from media coverage of green consumers playing a leading role in energy conservation.

The strong probability that some consumers will end up paying more for electricity under DR pricing has already stirred criticism of smart metering and variable pricing models from consumer advocate groups in the United States. According to an article in *Business Week*:

> From California to Colorado to Maine, consumer groups are expressing concerns about these efforts. In particular, they're leery of giving utilities the ability to change electricity prices on the fly, jacking rates up on hot summer days, for instance. Most utilities are prohibited from using variable prices now, but the flexibility to raise rates for a community as demand rises is essential for utilities to get the full benefit of new technology. Consumer groups worry these so-called smart-grid technologies are just another way for utilities to make extra money off consumers.
>
> The controversial aspect of smart meters is that they let utilities create real-time markets for electricity, with prices that could vary by the minute. If demand is low on a cool fall night, prices are low. But if demand surges during the summer as people blast the air conditioning, the utility can raise prices in the region to discourage usage. That saves utilities money because they wouldn't have to build power plants to meet high demand. And, supporters argue, consumers would learn to conserve energy. "Getting these different rate structures will help people understand the cost of energy," says William Gausman, senior vice-president for asset management and planning at Pepco Holdings Inc., a utility based in Washington, D.C.
>
> (Green, 2009)

As we have seen, the concern that DR pricing implementation will result in some customers paying more for the same amount of power consumption as before is well grounded. Proponents of DR pricing, such as the above-quoted William Gausman, argue that this higher pricing structure is required to motivate consumers to change their patterns of electricity usage. State regulators find themselves in the middle, trying to balance the arguments of utility companies for maintaining or increasing revenues and FERC's support of expanding DR programs on one hand with the potential for negative impacts of DR programs on cost of electricity for residential customers on the other. Many regulators have decided that utilities should demonstrate the impact of smart meters and variable pricing programs by setting up small-scale pilot projects, and then measuring and reporting on the results of variable pricing and smart metering on actual energy consumption and costs for residential customers before implementing a full-scale program of smart meter installation. The next section will review the results of selected consumer pilot programs and discuss the implications of these results for broader adoption of DR programs in the USA.

CONSUMER PILOT PROGRAMS AND RESULTS

Dozens of variable pricing and smart metering pilot projects have been completed in the past several years and many others are still underway. A study of pilot project outcomes authored by Nancy Brockway (Brockway, 2008) provides summary data and a detailed analysis of how residential customers have fared under such programs. The Brockway study reviews the results of three DR pricing programs in detail. These are the California Smart Pricing Pilot (SPP), which ran from 2003 to 2004, the Illinois Commonwealth Edison (ComEd) Smart Pricing Plan from 2003 to 2006, and the Ontario Smart Price pilot from 2006 to 2007. Each pilot program recruited volunteer households to participate in the use of variable pricing plans. Variable prices on the upside were triggered by utility-declared Critical Peak Pricing (CPP) periods during which the volunteers agreed to pay significantly higher rates for electric power. Participants were then charged at standard or substandard rates during normal or low demand periods depending on the design of the pilot.

According to Brockway: "All three pilots showed load shifts by residential customers on average in response to critical peak price signals" (Brockway, 2008: 41). However, Brockway reports that the average result data is not fully representative of consumer response to variable pricing because a minority of participants made very significant usage cuts during CPP periods and this minority accounted for most of the load shifts. In two of the pilots, some participants actually used more power during peak periods and others did not change their usage at all. These results lead Brockway to conclude that: "Not all customers (or groups of customers) will reduce their loads in response to higher peak prices. Indeed, customers may actually increase peak loads in any given peak period, despite the higher unit price they will pay for such usage" (Brockway, 2008: 43). In all three pilot programs analyzed, the CPP rates were between five and ten times the off-peak pricing rates, so the price differential was considerable.

Brockway notes that the small number of participants in each pilot, the volunteer nature of participation, and the variability of environmental conditions during CPP periods make it impossible to draw any statistically valid conclusions about whether pilot participant behavior was characteristic of the consumer population as a whole, or whether it would persist over time in the absence of the special attention given to the volunteers during pilot participation. In particular, she notes that the impact of DR pricing on average electricity bills for consumers cannot be predicted from the pilot project results collected to date, concluding:

> None of the pilots provides readily available information on
> likely bill impacts of AMI ... This omission is a major gap
> in the research to date, and hampers regulators trying to
> anticipate how an overall positive cost-benefit calculation
> for AMI will translate to specific customer groups. Findings
> of lowered bills from time-varying pilot prices must be
> discounted by the fact that the cost side of the equation

ignored AMI costs. Even without counting AMI costs, twenty percent or more of the CA SPP participants on all pilot rates saw higher bills. In the Ontario SPP, twenty-five percent of the participants had no bill decrease, or had bill increases on the time-varying tariffs.

*(Brockway, 2008: 83–84)*

This detailed analysis of the variability of consumer reductions in energy usage, from a minority of consumers who make very dramatic reductions to another group which makes none at all, contrasts with frequently quoted average results of pilot projects that cite overall average energy usage reductions on the order of 15 percent. Even consumers who did shift a significant portion of their energy usage to off-peak periods would need to commit to a permanent behavior change to lower their energy costs. It is exactly this critical outcome that analyses of long-term behavior of consumers participating in variable pricing programs indicate is unlikely.

Since smart metering and variable pricing pilot projects rely on consumers who participate voluntarily, and since pilot projects typically provide special attention and additional information for participants, it is likely that the general population of utility customers would not respond as positively to variable pricing and DR programs as did the pilot project participants. Unfortunately, this reinforces the perspective of the FERC assessment report that DR implementation may need to rely on mandated consumer participation in DR programs to achieve the full potential for reducing national energy consumption. The emphasis on DR strategies and controlling consumer consumption through smart meters, however, is only one side of the smart energy opportunity. More positive, but potentially even more disruptive for the utility industry, are programs that allow consumers to generate and manage their own power and to sell some of that power through the smart grid, a practice referred to as net metering. The next section will analyze alternative energy programs and products for consumers and discuss the barriers to widespread implementation of such programs that exist in the USA today.

## WHAT CAN CONSUMERS DO? OPTIONS FOR PERSONAL ENERGY MANAGEMENT

Based on the IBM Global Utility Consumer Survey of over 5,000 consumers in twelve countries around the world, more than half of consumers would be unlikely to change their energy consumption behavior significantly in a DR program without major financial incentives to do so. However, more than 20 percent of consumers are interested in taking charge of their personal energy consumption, a positive indicator for the future of smart energy management products for residential customers.

The IBM report characterized its consumer respondents using four energy consumption profiles. Passive Ratepayers, who make up 31 percent of those surveyed in 2008, expressed no interest in changing the status quo of their energy consumption. At the other end of the spectrum are the Energy Stalwarts, the 21 percent of consumers who are eager to take charge of energy consumption and even willing to invest in innovative solutions such as residential solar installations to achieve their goals. Allied with the Stalwarts in their level of interest, but less able to pay for improvements in their home energy infrastructure, are the Frugal Goal Seekers, the 22 percent of consumers whose main interest is in reducing the cost of energy usage. This group is the most highly motivated to change their energy consumption behavior in exchange for even a small reduction in monthly cost. The final group, 26 percent of consumers dubbed the Energy Epicures, keep up with information about the latest energy management options but express less willingness to make a change in behavior or an investment in innovative solutions to take control of their energy consumption. It would take large savings – defined in the survey as a 50 percent discount in consumers' utility bills – to motivate a significant change in behavior such as shifting their time of use (Valocchi *et al.*, 2008: 4).

Mapping these energy consumption profiles onto the type of variable pricing rate structure characteristic of the pilot projects analyzed by Brockway sheds an interesting light on how many residential consumers would be likely to change their energy usage. It seems highly likely that the Energy Stalwart and the Frugal Goal Seeker groups would respond to critical peak pricing by shifting their energy consumption to off-peak periods as much as possible. Consumers in these two groups would probably account for the dramatic load-shifting behavior documented in the pilot projects. Energy Epicures, on the other hand, may volunteer for a pilot project because of their desire for information about the latest developments but end up maintaining or even increasing their rates of energy usage after the novelty has passed and in spite of notifications that critical peak pricing is in effect. Since Passive Ratepayers are the least likely to volunteer for pilot projects, it is difficult to predict how they might respond if they found themselves in a mandatory DR program. Even if most of the Passive Ratepayers did change their consumption patterns under the pressure of higher peak pricing, some of them would probably maintain their existing energy usage behavior. In this extrapolation, between 20 and 30 percent of consumers would probably receive higher bills during critical peak pricing periods and fewer than 50 percent of consumers would be sufficiently motivated to permanently shift their patterns of energy consumption to reduce their costs.

As Brockway points out, pilot project results do not provide enough data to predict the impact of widespread DR pricing for residential customers. Consumer survey results are even more speculative when it comes to determining future energy consumption behavior. Nonetheless, the results of pilot projects and surveys underscore the lack of a clear consumer value proposition for smart metering as it is being rolled out by most utility companies in the USA.

Even with smart meters in place, consumers have relatively few options for managing their residential energy consumption costs. They can (1) cut back on consumption, (2) shift the times they use appliances, or (3) if their budget allows they can invest in residential energy-generation equipment. Consumers have control of the first two options, but as we have seen, under proposed DR program guidelines these changes in behavior would not result in much cost reduction. The third option, residential energy generation, is a more promising avenue for cost management that could have significant consumer uptake and impact. But this option is strictly controlled by the utility companies.

A further examination of each category shows that many smart energy products available to consumers fall into the first two options, which have limited cost-savings impacts. While the third option has the greatest potential for industry disruption and transformation, it is the one that utility companies have been reluctant to see expanded beyond a small group of customers.

(1) Cut back on energy consumption through substituting high-consumption appliances and other electric energy components for more efficient options

Whether from a desire to save money or to conserve energy, many action-oriented consumers have already changed their light bulbs from incandescent to fluorescent and LED options, insulated their homes, and adjusted their heating and cooling systems. The next stage of substitution involves updating appliances to more energy-efficient models, some of which include a wireless connection to the smart meter to respond automatically to DR pricing cycles. Appliance replacement has a high upfront cost and a small impact on annual electricity bills. Purchasing a smart energy appliance that costs hundreds of dollars just to save $25 or $50 in electricity costs each year may be feasible for a small number of affluent consumers, but it is not a realistic option for most consumers. Without strong government incentives or rebates, or a sharp increase in overall electric energy costs, upgrading of appliances on a large scale is unlikely.

(2) Monitor consumption and change behavior as needed to reduce electricity usage and reduce monthly bills

Another alternate is for consumers to monitor their use of electric devices to track which ones are most expensive to run at peak times and to shift the time of day. There are many low-cost and free smart product solutions to facilitate home energy consumption monitoring, including web-based energy tracking using free applications like Google's PowerMeter™ and Microsoft Hohm™. Some utility companies offer free in-home energy monitors to customers who agree to participate in DR programs, while others charge a monthly fee for such devices. As pilot project results have shown, many consumers are willing to change their behavior to save energy costs. However, there are limits to how much time shifting is feasible, and consumers who work from their homes, have large families, or medical needs may have little choice about the use of energy during peak demand periods.

(3) Create power for personal use and sell any excess back to the grid

According to the IBM survey, the 21 percent of consumers who are Energy Stalwarts are likely candidates for programs to generate residential power. However, the actual number of US consumers who have installed power-generation solutions is far smaller. From the consumer viewpoint, the high costs of purchase and installation of solar panels or wind turbines, the zoning restrictions in many neighborhoods, and the problems of maintenance combined with the long period required for any payback have been daunting barriers to residential power generation.

Supporting and motivating the desire of many consumers for home generation of renewable power would seem to be a natural step for utilities. In fact, most utilities are not yet open to unlimited net metering, with consumer participation and payback being capped or limited to credits on their use of electricity from the grid maintained by the utility company. Many utilities are reducing the incentives and rebates for solar installations and cutting back on new applications from consumers who want to connect a solar installation to the grid. As more consumers start generating their own power and using net metering to bring their monthly bill down to zero, utilities are looking for ways to impose new fees on solar power adopters rather than extending their programs to encourage even more active consumer participation in meeting renewable energy goals.

Consumer behavior in other countries demonstrates that with the appropriate motivations and incentives, consumers are eager to take an active role in residential energy generation. Consumers and small business owners have responded with particular enthusiasm to the prospect of long-term guaranteed payments for all the renewable energy that they generate. This type of renewable energy contracting is available in over forty countries, often with the extra incentive that payments for renewable energy are set by government policy at a price that is higher than the cost of traditionally generated energy. The combination of contracted long-term payments and premium pricing is characteristic of Feed-In Tariff (FIT) programs. According to the National Renewable Energy Laboratory (NREL):

> A feed-in tariff (FIT) is an energy-supply policy focused on supporting the development of new renewable power generation. In the United States, FIT policies may require utilities to purchase either electricity, or both electricity and the renewable energy (RE) attributes from eligible renewable energy generators. The FIT contract provides a guarantee of payments in dollars per kilowatt hour ($/kWh) for the full output of the system for a guaranteed period of time (typically 15–20 years). A separate meter is required to track the actual total system output.
>
> *(Cory et al., 2009)*

Germany's success in generating renewable energy and creating a fast-growing green energy business sector has been attributed to its support for FITs starting in the 1990s. The NREL report notes that there have been experimental FITs programs in the USA but that so far these have been quite limited by caps on project size, cost, and participation. Consumers filled all the slots in a FITs program in Florida just two days after applications for participation became available. Other states have also experienced this intense demand for slots in their programs. Unlike Germany and other countries with national FITs program development, it is up to individual utilities and states to determine the level of funding and support for any FITs initiatives in the USA. As a result, only a small fraction of interested consumers are able to participate.

Unlike FITs programs which pay cash for all the energy that a participant generates over the life of the contract, the net metering programs available from many US utilities are limited to providing consumers who generate residential electricity with a credit against their bill. If the power generated exceeds the amount coming from the grid, the consumer's monthly bill goes down to zero, but no payment is made for excess power generation. Even this limited incentive has created more interest from US consumers than utility companies are willing or able to meet.

For example, Xcel Energy in Colorado proposed a new monthly fee for solar customers in the summer of 2009. Xcel justified the new fee structure as a way to force consumers who routinely contributed more energy than they used (and therefore did not pay anything to the utility) to contribute to Xcel's overall grid capacity costs. Xcel's claim was that generation capacity was being kept in reserve for these customers just in case they needed to draw power from the grid at some future date. An Xcel spokesman explained that even net solar contributors had an obligation to pay for the whole infrastructure just like customers who depended on Xcel for all of their power needs. If they didn't pick up their fair share of the infrastructure costs the solar customers would in fact be getting an unfair "windfall" return from their investment in installing and maintaining solar panels. The proposed monthly fee was small, but the precedent of imposing new charges on consumers who had already paid a considerable amount to buy and connect their residential solar installations led to widespread protests around the state. In the face of political and consumer opposition, Xcel withdrew its monthly fee proposal for the short term but reiterated that it would be back with another version before long, calling the increase in solar customers a "growing issue" which had to be addressed (Proctor, 2009).

The utility industry's insistence on control, fees, and limits on net metering is especially striking since only a tiny fraction of the residential utility customers in the USA have installed the solar, wind, or other renewable energy resources that qualify them to participate in net metering programs. The government data published in April 2009 report that a total of 48,820 utility customers in the entire USA were participating in net metering programs at the end of 2007. Of these, 72 percent were in California, the largest state by far in terms of solar power adoption. The fact that these numbers are increasing year over year seems to be a source of alarm to utility companies rather than a cause for celebration. The total number of solar residential customers in Colorado as of June 2009 was 5,661 and of these about 3,000 were generating enough electricity to reduce their utility bills to zero, thus causing a "growing issue" for Xcel Energy's revenue projections.

Colorado has a total population of 4.8 million, with over 1.6 million households that are Xcel customers. Xcel Energy reported total 2008 revenues of $11.2 billion and operating profits of $646 million (Xcel Energy, 2009). It has invested over $100 million in its heavily publicized "Smart Grid City" project in Boulder, Colorado, which it promotes as "the world's first fully functioning smart grid enabled city." If Xcel Energy considers 3,000 net metering customers to be a major issue requiring changes to its rate structure, it is hard to imagine that it is prepared to support any meaningful percentage of its residential customer base becoming active energy generators, much less the estimated 21 percent of Energy Stalwarts who expressed a desire to take control of their energy future.

Roger Duncan, the General Manager of Austin Energy, another utility company that publicizes its support of consumer solar installations and smart metering, says candidly that utilities are simply not prepared to deal with large numbers of customers generating their own energy:

> Given the structure of our high solar rebates and a net metering policy where Austin Energy buys back power from its solar customers at full retail cost, we would go bankrupt if too many customers signed up for these programs. The regulatory environment is not set up to deal with large-scale residential power generation and activist consumers, neither is the business model of utilities. Smart meters highlight fundamental issues that are going to require fundamental changes in the way we think about the role of consumers. Utility companies just haven't made that shift yet.
>
> *(Duncan, 2009)*

Austin Energy was one of the first utilities in the USA to install smart meters, upgrading the meter infrastructure for all of its 400,000-plus residential customers during 2009. Its smart metering proposal highlighted the ways that this new and improved infrastructure would facilitate consumer participation in energy management and power generation. Despite considering itself a national leader, Austin has a long way to go – as of September 2009 there were only 900 residential solar rooftop installations in the city. But pent-up consumer demand has run into the barrier of Austin Energy's business concerns. With over 400 pending applications for new residential solar permits, Austin has begun limiting the growth of solar customers. It announced a hiatus in accepting new applications and reduced its incentive rebate program and net metering rates in fall 2009 to control the financial impact of too many active power-generating consumers (Austin Energy, 2009).

If Xcel and Austin Energy are unable to cope with more than a small percentage of their consumer base taking advantage of solar incentive programs, there is little reason to believe that US utility companies in general will see their investment in smart meter installation as a way to encourage an increase in residential energy generation. This leaves consumers with the frustrating prospect of paying a surcharge for a smart energy infrastructure that doesn't provide them with any new options for controlling their own energy resources or even managing their residential energy costs. Rather than giving customers smarter solutions for energy management, the smart grid initiative to date is preserving the long-standing Hegemon control model of the utility companies.

## THE HEGEMON ECOSYSTEM AS A BARRIER
## TO SMART ENERGY SOLUTIONS

After more than a hundred years of incremental improvements, electric meters and power grids are making a leap into the future with connected, embedded intelligence. Consumer interest in active energy management is increasing rapidly – too fast for utility companies to keep pace. Passive Ratepayers are now in the minority among residential utility customers and consumer interest in renewable energy ranks second only to their desire to manage energy usage to control costs. In a 2009 survey, over 90 percent of US utility customers said that they were concerned with residential energy costs and interested in receiving detailed information on their energy use. In the same survey 76 percent expressed an interest in renewable energy technologies for their home, and 72 percent of those respondents think that "reducing personal energy costs" is the most important benefit of renewable energy (Oracle, 2009). Another 2009 consumer research study reported that 43 percent of US consumers would consider installing a solar panel in their home (CSA International, 2009).

Utility companies give the impression of keeping up with these new consumer attitudes as they forge ahead with smart meter installation and smart grid projects. A closer look, however, reveals that most utility company business models and smart grid strategies are not prepared for more than a token number of activist energy-generating customers. When the limited number of application slots for consumer net metering and renewable energy rebates are taken, utilities quickly shut the door and tell disappointed customers to try again next year. Behind their green exteriors, the power that utility companies have historically wielded over their markets and their ecosystem partners continues to function very much as it did in the smokestack days.

Utility companies prefer Hegemon ecosystem models that place them firmly in control of all energy-related activities and allow them to adopt new technologies at a slow, controlled, and incremental pace. Since the utilities themselves have a track record of specifying and building highly customized plant additions rather than creating interoperable technology platforms, the competition to become a Federator in this rich and growing industry sector is fierce. Questions about what standards will be accepted for key infrastructure components and who will build the technology platforms that are needed to serve as the center of a smart grid federated ecosystem are still very much unanswered. Large and small technology companies are lining up to participate in smart meter pilot projects, providing various solutions and services that they hope will be adopted by utility customers. Rather than striving to disrupt utility business models with transformational technology, however, smart solution providers from Google and Microsoft to small, innovative start-ups seem intent on gaining admission to the Hegemon-controlled utility ecosystems.

Pepco is one of dozens of utilities in the USA that have attracted a flock of pilot project partners. Its "PowerCentsDC™" pilot project with variable pricing, smart meters, and smart thermostats started in July 2008 in Washington, DC and includes a mere 1,400 volunteer residential customers. The long list of companies and organizations involved in designing and managing the pilot project for these residents includes the Smart Meter Pilot Program, Inc., a non-profit corporation consisting of the District of Columbia Public Service Commission, Office of People's Counsel, Consumers Utility Board, International Brotherhood of Electrical Workers, and Pepco Holdings, Inc. along with UtiliPoint® International, Inc. providing project management and rate design services, eMeter providing consulting services, Sensus providing the smart meters and wireless infrastructure for two-way data exchange, Mincom measuring usage in the different variable rate structures and providing billing support, and Comverge providing smart thermostats to selected residents (UtiliPoint, 2008).

The crowd of companies taking part in the PowerCentsDC pilot project reflects the pervasive rush to join utility company ecosystems and stake a claim in the smart energy sector. Attracted by the billions of dollars that will be spent over the next decade on energy infrastructure, equipment, and information technology everyone from multinational conglomerates to garage start-ups is competing for a piece of the smart grid action. Venture capitalists are investing in scores of small companies that have developed innovative smart energy technology in the hopes of creating the next Microsoft or Google, Oracle or Cisco. And these original tech giants are promoting their own smart energy solutions and establishing new consortia to promote them.

Consortia and industry associations are springing up at a brisk pace to shape technology solutions for different high-growth sectors. In addition to the long-standing groups such as Gridwise and EPRI, dozens of newly formed energy groups are contending over technology platforms and standards. The sheer number of new smart grid forums, alliances, conferences, and associations that are working on creating standards, specifications, solutions, and platforms has added to the confusion of a sector that was already struggling with a lack of interoperability among various smart grid components.

To define a core set of standards for smart grid development, the National Institute of Standards and Technology (NIST) contracted with EPRI in 2009 to develop a model for an overarching smart grid architecture and to create a framework for the interoperability of protocols and standards for smart grid systems and devices. The majority of this interoperability effort focuses on power generation, transmission, regional distribution, and security, and not inside the home where consumers are looking for information about energy usage and cost control.

A draft report published in fall 2009 identified more than eighty specific standards and specifications that were related to smart grid infrastructure, and called out a number of gaps where additional or revised standards would be needed, including plug-in electric vehicles and home energy management systems. These missing standards will not be available until the second half of 2010 or later (National Institute of Standards and Technology, 2009). In the interim, companies with competitive smart grid and home energy management products continue to argue over basic issues such as whether to use Internet protocols, a short-range wireless protocol like ZigBee, or proprietary networks for connecting components and transmitting data between smart meters and utilities and between smart meters and home appliances and energy monitoring devices.

One strong proponent for standardization of the interface between smart meters and residential energy information solutions is the ZigBee Alliance. The ZigBee protocol supports internal wireless communications between the smart meter and a home network, smart thermostat, or smart appliance interface. This allows signals from the utility company regarding peak pricing periods or critical demand situations to be passed on to other ZigBee-enabled devices along with instructions about how to respond. If the consumer in a DR program has a smart thermostat, for example, the utility will transmit notice through the smart meter that participating DR customers need to reduce power consumption immediately and keep consumption low for a specified period. The meter sends appropriate signals to the smart thermostat which then cycles off the air conditioning for the required period. If other appliances such as clothes dryers or swimming pool pumps are attached to the smart meter, they will also be put onto a reduced power diet.

Companies that make smart devices with ZigBee capability see the smart meter as a platform that can become the center of a Federated ecosystem of interconnected in-home devices. But in the United States, the actual capabilities of installed smart meters depend largely on the priorities determined by local or state utilities that provision and configure the meters and select the vendors to implement the smart meter program. Like the landline and wireless carriers discussed in Chapter 5, utility companies have a long history of controlling all aspects of their ecosystems, including any products or services that are delivered under their auspices. And as we have seen, the utility companies in the USA are in a position to slow down the implementation of any consumer-controlled energy management options that threaten to disrupt their existing business models.

CONCLUSION

Utility companies, industry regulators, and government planners publicly agree that smart meters will play a critical role in the emerging smart grid ecosystem but there is no shared vision about what that role will be. Smart meters are expected to create opportunities and incentives for consumers to participate in energy conservation, generation and management as well as delivering significant benefits to utility companies. Unquestionably, however, the most immediate beneficiary of smart meter installation is the utility company rather than the consumer. Residential customers may benefit indirectly from improved utility company efficiency, but they are several steps removed from seeing any tangible, personal benefits. In particular, the opportunities for saving money on their electric bills, the highest priority for most consumers, are embedded in complex regulations and calculations condensed down into rate plans and peak demand pricing models.

Once smart meters and a smart grid are in place, residential consumers may have an opportunity to access much more detailed information about their electricity consumption. But putting this data to work will require that they buy and connect other smart products to their meters or home monitoring systems. As consumers participating in DR pilot projects have discovered, usage information does not necessarily result in any reduction in monthly electric bills. To obtain cost reduction, or perhaps just to avoid the threat of a significant cost increase, consumers must also be willing and able to change the way that they use electricity on a day-to-day or even hour-to-hour basis. Opportunities to assume more direct control by installing renewable energy such as residential solar or wind generation are still expensive and limited to a very small minority of early adopters.

The utilities are still exercising Hegemon control over smart meter interconnections in a manner quite reminiscent of the telecommunication Hegemons in the pre-Carterfone era. Grid connections are subject to review, approval, and fees regardless of whether the connection is to a large residential solar system or a read-only device connected to the meter that tells consumers about the details of their hour-by-hour electricity consumption.

The still-to-be explored capabilities of a smart grid and a smart meter infrastructure, the mandates for the adoption of renewable energy technologies, and the role of national and regional governments in providing incentives for consumer generation of power may one day combine to create new business models for energy generation and distribution and new smart energy services. This combination has not yet reached anywhere near a level of maturity that might disrupt the utility Hegemon's degree of control or way of doing business in the USA.

Until smart energy options for consumers expand from simple usage monitoring and reacting to DR pricing signals to control over residential energy generation, any transformational impact of smart meter installation will remain camped just outside the consumer's front door.

# 7    Smart home vision and reality

Business plan competitions are popular again in the USA as are incubator programs for early stage start-up companies. To help participants prepare for the rigors of real-world meetings with venture capital investors, incubators often feature a session where fledgling companies can practice and perfect their fund-raising pitch. The theory is that start-up teams will benefit from the candid assessment and advice of the seasoned industry veterans who typically serve as mentors and judges at these events.

In the spring of 2009 a start-up management team took the floor at one such incubator to present its plan for launching a home networking company that would integrate various smart entertainment devices, appliances, energy monitors, and PCs. The team presented ample evidence of unmet market demand and unsolved technical challenges in current home networking platforms. Instead of friendly advice, however, this team received a chorus of negative feedback about its plans. According to the panel of mentors the home networking space was a guaranteed death trap for any small technology company. The skepticism was so pervasive that management eventually shifted the company's focus to a product with a different value proposition. In aspiring to become the lynchpin of home networking, however, the start-up team was undeniably following in some well-trodden and notable technical footsteps.

Samsung trademarked the phrase "Home Wide Web" (HWW) in 1998 for its widely publicized technology platform for connecting PCs, TV sets, remote controls, modems, VCRs, and cable boxes using a standard called IP (Internet Protocol) over IEEE 1394 (Samsung, 1998). Despite the catchy HWW nomenclature and

Samsung's attempts to create a Federator ecosystem, including an industry-wide alliance to support implementation of IP over IEEE 1394 in all digital consumer electronics, the HWW platform never attracted enough ecosystem partners to make a lasting market impact.

From a different technology platform perspective, Lucent Technologies partnered with a company called Tut Systems in 1998 to develop Home Wire chips that would allow consumers to create a home network using the standard copper phone wiring that was already in place for the then ubiquitous copper wire landlines. Lucent's positioning for Home Wire™ was that the "phone line is the 'quick-strike capability' that's affordable and simple enough to spur consumers to install home networks" (CNET News, 1998). That approach to jump-starting home networking also failed to capture any significant market share.

At the beginning of 1999 Sun Microsystems formally released its Java-based Jini networking technology as a home networking solution that promised to "connect-anything-to-everything" inside and outside the consumer home. The Sun rollout program for Jini featured demonstrations of using Jini to connect devices like digital cameras, printers, dishwashers, and Palm Pilots (Plamondon, 1999).

Perhaps fearing it would miss the train, that same week in January 1999 Cisco announced that it too had plans to tailor its enterprise local area networking technologies to serve as a technology platform for home and consumer device networks. Bethany Mayer, Senior Product Manager for Cisco, elaborated: "The plan and strategy is to move multiple devices into the home – like a LAN for the home. You could network so many things – IP phones, all your alarm systems, you could network fridges, or Web devices, and PCs" (Miles, 1999).

Finally, not to be outdone in the hotly contested race of January 1999 to stake a claim as the home networking platform of choice, Microsoft also announced its own networking standard to connect smart electronic devices. Microsoft called its solution Universal Plug and Play™. According to Craig Mundie, Senior Vice President of Consumer Strategy for Microsoft at the time: "We want to bridge the gap between appliances, products, and computers." As the coverage in CNET News duly reported: "Microsoft wants to easily connect existing intelligent devices such as computers to a growing class of what Mundie termed 'smart objects'" (Davis and Kanellos, 1999).

Observing these market-leading companies and a cluster of home networking start-ups making related technology announcements, an analyst for International Data Corporation spotted a trend in the making: "There is a market out there for networking in the home," but astutely held back from predicting exactly when home networks would become a commonplace reality. "While certainly there's some real opportunity in 1999, in terms of a mass market, we're talking years" (Heskett, 1999). Even that cautious analyst forecast turned out to be optimistic. When it comes to connecting everything-to-anything in the home on a common network, perhaps he should have said, "we're talking decades." We certainly are not there yet.

Having presided over the demise of companies that they funded during the 1999 wave of home networking solutions, and having endured the grim reality of personally trying to connect devices from different vendors into any of the available home networking platforms, the veteran mentors judging start-up pitches for home networks can perhaps be forgiven for a touch of skepticism about the chances of success for yet another new entrant in the connected home arena.

It's not that the need for seamless home connectivity has been resolved. The average consumer has added dozens of smart devices to their home environment since 1999 and discovered that getting those devices to communicate with each other has become ever more complex. Smart products routinely send data to – and receive data from – the world outside the home, but each device embodies the protocols and network requirements of its vendor, which are likely to be quite different from those of another company's product sitting a few feet away. That very complexity of organizing smart devices with so many types of embedded connectivity options has been a frequent source of frustration for individual consumers and solution providers alike. It may be a strange inversion of Metcalfe's Law that the more devices that are added to the home, the less communication there is between these devices.

One benchmark example is the long-time quest for a truly simple, easy to set up, universal remote control that works out of the box with all the different types of smart, connected products in today's home. Just in the home entertainment space, the number of separate modules has mushroomed along with the ingenuity required to link some or all of these boxes to a single controller. The cast of characters includes HD television displays; game boxes from Nintendo, Sony, and Microsoft; branded music players like iPods; HD radio receivers from Bose and Livio; HD-DVD players from electronics manufacturers; broadcast recording devices from venerable VCRs to DVRs and TiVo™ recorders; streaming media devices such as Roku™ and Boxee™ and, to top it all off, the cable box or satellite TV receiver. Every one of these devices arrives with its own codes and operating behaviors all neatly encapsulated and made manifest in its very own remote control. The perpetual challenge to home consumers who don't want to take on the role of network administrators is reflected in this statistic from Logitech, one of many vendors of universal remotes, as part of its instructions for activating the high-end Harmony controller: "Your Harmony supports more than 225,000 devices from 5,000 brands" (Logitech, 2009). The statement is probably intended to provide reassurance to customers that no matter how wildly enthusiastic they are about collecting digital entertainment devices, they can count on Logitech to cover everything they own. The impact on non-technical consumers is more likely to be closer to despair about ever getting control of the unruly herd of smart devices that have taken over the entertainment room.

A sense of home networking skepticism among consumers permeates the results of the 2008 State of the Connected Home Market Study that was commissioned by the Continental Automated Buildings Association (CABA). This study is the third in a series of reports based on CABA surveys of consumer attitudes toward home automation and connectivity. Since the prior study in 2005, the number of US online households that responded with "strong interest" to "Whole Home Control" (characterized in the survey as networks connecting home entertainment, energy management, and security systems) has remained stubbornly low at about 5 percent of respondents. Even though over 90 percent of US respondents reported that they used a computer for entertainment in the previous three months, only about one-third described themselves as "very interested" in connected entertainment ecosystems. This was a decrease from the two-fifths who were "very interested" in the 2005 survey (CABA, 2008). Perhaps the 2008 respondents had in the interim actually attempted to connect their PC to their TV to their DVR to their game box of choice.

Despite the marked lack of consumer enthusiasm, at least some of the sponsors of the Connected Home Market Study managed to interpret the 2008 survey results as an endorsement for even more home networking solutions – with the caveat that the connections had to be simple: "What consumers want most is an easy, seamless way to integrate their smart home devices – their mobile device, their TV, their appliances, you name it:" concluded Carol Priefert, Senior Manager, Whirlpool Corporation, in a statement for the CABA press release about the study. And Tony Wan, Director of Marketing at Cisco, commented: "As highlighted by this market study, the proliferation of IP devices in the home and increasing consumption of digital media creates abundant opportunities for the broader connected home ecosystem to work together to deliver the seamless experience that consumers are demanding" (CABA, 2009).

It is understandable that networking technology companies and appliance makers would want to put a positive spin on the survey data, especially when they have already made significant technology investments in embedded product connectivity and home networking solutions. Nonetheless, the results of the 2008 State of the Connected Home Market Study are an important reminder that the integration of home networks and seamless convergence of connected smart products that solution providers have been championing since 1999 is not a top priority for consumers. In fact, there are good reasons to resist the urge to connect everything-to-anything inside the smart home. In addition to avoiding some short-term frustrations inherent in working with non-compatible device interfaces, keeping their home product spheres separated is one way for consumers to maintain some personal control over increasingly autonomous product connections and communications.

This chapter reviews scenarios that selected vendors have presented during the past decade to market their vision of the smart home of the future. It analyzes the consumer value proposition and ecosystem models implicit in this vision, discusses why centralized home networking has not materialized, and presents a scenario for developing home smart ecosystems. Emerging sectors such as home energy and health monitoring are considered as potential models for smart services in the coming decade. The chapter concludes that continued competition among industries and technology platforms for a role in home networking has on balance been a positive factor in the mass-market adoption of smart consumer products.

SMART HOMES OF THE FUTURE: VENDOR SCENARIOS
AND CONSUMER REALITIES

Here's how the home of the future looked through the eyes of Microsoft in 2003:

Instead of traditional locks, there's an electronic kiosk with
a touchscreen, a biometric scanner, and a smartcard reader.
Go ahead and make eye contact; if you're a match, you'll pass
through into your future home – a time and place a half-dozen
years from now when your living quarters will recognize you,
communicate with you, and anticipate your every need ... The
lights and heat automatically fine-tune to your preference the
moment you cross the threshold. A screen on the wall in the
foyer reads your email aloud as you hang your coat. Your kitchen
has become your own private sous chef. Run a chicken pot pie
beneath the barcode reader on the microwave and it sets the time
and temperature. Break out the food processor and some baking
material; your home recognizes RFID tags in the bag of flour and
offers to help.

*(Wired, 2003)*

In Korea, the home of the future as described in 2004 features broadband connectivity for every room and for products used in all aspects
of consumer life. In addition to espousing a vision of smart, connected kitchen appliances, video-enhanced security and responsive
entertainment systems, Korean vendors were early promoters of
home health monitoring:

The South Korean government and several larger companies
here are actively promoting technology for connecting every
household appliance to the Internet ... Samsung's own demo
home has a bathroom health monitor that enables you to send
information on vital signs such as blood pressure and oxygen
levels to your doctor while you sit on the toilet. The home
network also includes Internet-activated rice cookers, screens
in the bedroom that let you see who is at the front door, and TV
sets that can flash up the number of an incoming phone call.

*(Kanellos, 2004)*

In 2000, Nokia's vision of the connected home of the future featured
an integrated media, news, and entertainment portal that would

support wireless home shopping, video gaming, and full movie access as well as a mobile phone interface:

> Using a mobile device such as a WAP phone or a mobile internet display with Wireless LAN, the consumer can access the home network easily and download digital photos, make additions to common shopping lists and monitor the security system ... The Media Terminal enables full convergence of digital TV and Internet services, featuring a new kind of Media Portal. Media portals will offer interactive magazines with high-speed and perfect quality video clips, networked gaming and instant Internet access. It will also enable interactive advertising and shopping (Click & Buy); movies on demand (downloaded on local hard disc). The Media Terminal is the point in the home where all entertainment and information will meet and create new services.
>
> *(Nokia, 2000)*

More recently, the HomeLab concept home described online at Philips Research's website in 2009 features smart products that "range from electronics that recognize your voice and movement to digital displays within the bathroom mirror to new toys that will help children expand their creativity." Philips describes the future of smart home entertainment in these terms:

> Picture yourself relaxing at home on your couch. You're unwinding from a long day and want to play some music but you're too exhausted to move. Instead, you say "Music, where are you?" and hum your favorite slow tune. Lucky for you, your smart home entertainment system understands your needs. Not only does it play the song you were humming, it dims the lights to provide a more relaxing environment for you.
>
> *(Philips Research, 2009)*

Clearly there is no shortage of vision from appliance makers, electronics and entertainment brands, or IT and networking providers when it comes to imagining opportunities for smart, connected homes. Ironically, this plethora of ideas about the ideal future home has become one of many barriers to adoption of today's connected home solutions. Look a little closer, and it is obvious that each brand sees its own platform and services as the hub of the home. It's not a surprise that companies like Cisco, with deep expertise and product lines in routers and firewalls, see Internet connectivity as the starting point of all home network installations. Nor is it hard to understand why Microsoft frames the home of the future in terms of an integrated operating system that can deliver email and entertainment to distributed screens and workstations, or why Nokia sees it as a very large smartphone terminal, or why Philips features the comforts of smart home entertainment systems. Each vendor is promoting its own technology as the core standard to which all the other devices in the home will have to conform in order to connect. But what's in it for the consumer?

Some common threads are notable in the types of devices described by home networking providers in the 1999 announcements and in the future home scenarios quoted here. There are multiple mentions of video security systems, Internet-connected kitchen appliances, lighting controls, and personalized, distributed access to all forms of media and entertainment content. This triad of networked home entertainment, energy management, and security systems was also used in the CABA survey to define "Whole Home Control." Technology vendors are still focused on connecting this particular cluster of products despite a decade of lackluster home automation adoption by consumers and a lack of any strong interest on the part of the 2008 CABA survey respondents. Consumers are sending a clear message through surveys and their consistent lack of adoption of costly whole home networking solutions that certain types of connectivity are not that interesting or valuable to them. For today's consumer, home networking means the Internet and possibly a Wi-Fi access point. But that is a message that technology vendors do not want to hear.

The technology required to implement most of the features of smart homes described by vendors has already been commercialized and adopted for use in both commercial and residential settings. Biometric screening, contactless smart cards, and facial recognition, for example, are used in airport security systems and for access to high-security office and government facilities. Video surveillance and screening of visitors to offices, homes, and apartment buildings are common, as are caller ID systems for all types of phones. Luxury hotels are equipped to customize their room lighting, music, and entertainment systems to match the tastes of their regular guests and similar personalization is available in high-end homes with home automation systems. Despite the challenges of setting them up, connected home entertainment systems with universal controls are popular with consumers, propelled by the widespread adoption of flat-screen and high-definition television sets.

Other components of the future home scenarios favored by vendors are readily available to consumers but have not been widely adopted. While home appliances with embedded intelligence and Internet connections have been on the market for over a decade, bar-coded cooking instructions and RFID-activated recipe suggestions are harder to find outside of the laboratory. Smart, connected toilet seats with sensors to monitor various health indicators are, however, readily available to interested consumers in Korea and Japan.

It seems reasonable to conclude that unmet technology requirements are not the main barrier to smarter homes becoming a consumer reality. Promoting existing technology and making incremental changes to current products are not likely to increase consumer demand for proprietary home networking and smart home solutions. The barrage of competing announcements and technology platforms that elbowed one another in 1999 made it more difficult to build consensus for a standard platform for basic home networking functions. Every technology company entering the sector set out to win adherents to its own technology, to become the platform which all other devices had to connect to and communicate through. As in other industries, the competition between would-be Federators with incompatible platforms splintered the nascent market and reduced the potential for economies of scale and new common services and applications.

As more products are endowed with embedded connectivity, every new smart product that a consumer buys, from coffee makers to energy monitors and HD TV screens, adds another layer of complexity to networking requirements. Each of the dozens of connected products is designed to function optimally on its own channel, or in collaboration with a designated group of like-minded ecosystem partners. Convincing the manufacturers of these products to integrate across network protocols with new Federator platforms and potentially competing devices typically produces a cacophony instead of a soothing, seamless symphony of smart services.

Each vendor sees itself as the centerpiece and hub of home intelligence, connectivity, and smart services. Unfortunately this results in endless cycles of competition among multiple connected platform contenders rather than a manageable number of options. Global brands are typically determined to guard their brand identity and existing customer relationships. This makes it challenging for one company to establish a federated ecosystem that is completely integrated across the different industry sectors. Even when consortia are formed, there will always be holdouts and competitors with a platform that conforms to their notion of what should be the standard. When those competitors are major global corporations like Microsoft, Philips, and Samsung the home consumer is often forced to choose one or another brand and live within the extent of that company's ecosystem connections for a certain set of products and services.

## BUSINESS BARRIERS TO CONNECTING HOME MEDIA

Rather than buying new networking solutions, consumers would like it to be easier to simply connect their smart products to each other in configurations of their own making. Unfortunately, the devices they would most like to connect – home media and enter-tainment products – are the ones that are often designed with delib-erate, embedded barriers to connectivity. The business models of entertainment and media devices (TV, DVR, and DVD players, video game boxes, and digital music players) are based on broadcast advertising or payment to access and download media content. From the content owner's perspective, letting that digital content travel freely throughout the home and onto a home computer is a business model nightmare. Once stored on an Internet-connected computer, it is harder for the content owner to prevent the user from copy-ing content onto other devices, sharing it with friends, or making it available on the World Wide Web. While digital rights managements (DRM) solutions offer some anti-copy protections for digital content, most such content protections can be circumvented by a determined consumer with advice from Internet postings. The combination of proprietary formats and hardware-enforced restrictions which lock proprietary content to the vendor's proprietary player is much harder to break. Broadcasters, cable companies, TV, film and music produ-cers, and content owners have a strong interest in keeping their con-tent restricted to the specific products that conform to the rules of their controlled ecosystems. Moving that content onto a variety of Internet-connected devices outside of their control is a threat, not a benefit.

So far the barriers to moving media from one dedicated enter-tainment system to another have deterred Federator business models from taking root in the home but have not prevented determined consumers from making ad hoc connections to create customized home networks. Media companies are always looking for new ways to block such ad hoc connections, however. Attempts to embed DRM controls into entertainment devices such as HD TVs are heavily sup-ported by the trade groups such as the Motion Picture Association of America (MPAA) in the USA and the British Broadcasting Company (BBC) in the UK. In 2009 the MPAA began lobbying the Federal Communications Commission (FCC) to reconsider the ban on

Selectable Output Control (SOC), a method that could be used to disable all non-secure devices with direct connections to HD TV sets. Consumer advocates have successfully opposed SOC to date on the grounds that it would remove functionality from devices that consumers had already bought and installed, such as certain DVD players and sound systems. Even the MPAA has admitted that millions of legitimate home entertainment devices would be rendered useless by SOC, but it argues that enhanced broadcast security, by limiting access to approved digital-only devices, is needed. Without stronger hardware-based anti-copying controls, they claim, recent films cannot be made available via broadcast TV without danger of piracy. In the UK, the BBC revived a request for embedded DRM controls on the broadcast of TV programming, using an algorithm that would only be available to approved media and broadcast companies. The control-oriented mindset of content owners, media producers, and broadcast companies is a barrier to integrated home entertainment networks that is proving even tougher to overcome than the battle over networking platform technologies in the home (Kwun, 2008; O'Brien, 2009).

The lack of clear consumer value is another, more fundamental, business barrier to whole home networks. By itself, a network is not a very compelling value proposition even when it actually delivers on the seamless, simple, and plug-in everything-to-anything promises that have been made for the past decade. Sometimes connecting devices to each other just isn't all that useful from a consumer perspective. Unless companies are prepared to create smart services offerings based on integrated connected product ecosystems the cost and effort of making such connections will remain a barrier to consumer adoption. Nonetheless, developing solutions for the connected home still seems to hold a perennial and almost irresistible attraction for networking technology vendors and start-up companies. Industry veterans and new entrants continue to compete for a lead role in the smart home of the future, even in somewhat unlikely spots such as the kitchen.

## A SMARTER HOME ECOSYSTEM SCENARIO

DRM and business barriers to connecting media and entertainment devices may explain the fascination that so many technology vendors have with connecting kitchen appliances. The contents of your refrigerator may not be that entertaining, but at least they are not copyrighted and they clearly belong to you. But to create a compelling consumer value proposition, vendors need to offer more than a network connection to a cold box. Despite the starring role of the kitchen in many future home scenarios, the basic Internet-connected refrigerator is not much more appealing to consumers in 2010 than it was when it first appeared in the 1990s.

Is the connected kitchen appliance just a vendor fantasy, or is it a smart product in search of a supporting smart ecosystem? This section will consider what it would take to create a winning ecosystem that would transform the kitchen into a hub of connected activity, healthier living, energy conservation, and convenience. Networking the traditional refrigerator would be just the first step.

Imagine planning for a smart product that offers consumers innovative features, either by enhancing the capabilities of an already familiar device or by creating a novel alternative to a traditional product's feature set. To deliver on its full potential and to win customers, this smart new offering must overcome a number of barriers to market adoption. The more innovative the product's features, the more challenging it will be to ensure a smooth integration with complementary products and services. The infrastructure required for extending consumer applications and creating an optimal value proposition is not in place in today's homes. Previously adopted content, peripherals, and plug-in enhancements may not be interoperable with the new product, forcing consumers to upgrade or abandon current favorites when they buy the next smart product. The new product needs a smart ecosystem strategy that will deliver complementary services and enable customers to put its embedded intelligence to use immediately.

For example, suppose our hypothetical new product (SmartR, for "smart refrigerator"), aspires to revolutionize the way consumers shop, eat, and cook by becoming the intelligent hub for the home kitchen. SmartR will succeed by connecting multiple smart products in the home to create a bundle of related value-added services and consumer options. Dozens of service provider partners and hundreds of application developers must be motivated to join the SmartR ecosystem before launch to ensure that core services are available as soon as the product hits the market. Interconnected ecosystem components will help demonstrate the SmartR value proposition and convince consumers to buy it. Features like tracking expiration dates of perishable foods and sending shopping list reminders to SmartR owners would be just the beginning. A smart ecosystem would need to enable a variety of value-added services to make SmartR the brain center of household shopping, cooking, nutrition, and health-related activities. Which of the ecosystem models described in Chapter 2 will work best for the SmartR company?

If the SmartR company is a large and already well-established appliance manufacturer with the resources to develop advanced, proprietary smart products under its brand name, it might be inclined to adopt the Hegemon model for its market entrance strategy. One advantage of this approach is that the company creating the ecosystem maximizes its control of the features and functionality of all SmartR-related products and services. Current value chain relationships and sales channels are maintained and gradually expanded to support a broader ecosystem of services and content options. No disruption is required. In fact, this is the path that the first wave of smart appliance makers decided to take when developing and marketing Internet-connected kitchen appliances in the past decade.

LG Electronics, for example, introduced an Internet-enabled refrigerator in Korea in 2000 and in the USA in 2002. According to an LG press release announcing its US launch:

> The 26-cubic-foot refrigerator features a high-quality 15.1-inch TFT-LCD and its own LAN port to enable Internet surfing and shopping, as well as two-way videophone calls with friends

and family. It can also be used for watching TV, listening to music and e-mailing messages to friends. In addition, a digital camera mounted on top of the LCD enables video messages and digital still photos to be created, exchanged and printed out. The refrigerator makes extensive use of touch screens, a simplified graphics user interface, electronic pen and voice messaging for a user-friendly experience. Using these tools, consumers can check real-time price information on groceries; obtain tips on food, nutrition and recipes; be reminded of scheduled events; be informed when to change the refrigerator's filter, and learn cooking methods for products stored inside.

*(LG Electronics, 2002)*

By any standard, this is an impressively intelligent refrigerator. It has so many media and communication options that storing foods at the right temperature almost seems like an afterthought. Nonetheless, LG did have some innovative ideas for adding value to the traditional functions of the refrigerator. The potential for checking on grocery prices before shopping would be appealing to cost-conscious consumers, as long as the necessary information was available from local grocery stores. Reminders about changing the filter, on the other hand, would probably not motivate too many buyers. Keeping track of all the food stored in the refrigerator and using embedded intelligence to create shopping reminders and expiration date alerts are all expected attributes of a smart refrigerator. But LG's refrigerator wasn't quite smart enough to deliver this service automatically. Instead of a system that allowed quick scanning of each item's barcode label as it was put away, the owner had to create a food database by manually entering product names and expiration dates using the refrigerator's UI panel, and then remember to make a follow-up entry whenever items were consumed. Rather than creating an open specification and interoperable platform with a Federator model, to encourage ecosystem partners to integrate their applications and services with the smart refrigerator, LG built its product ecosystem from the inside out, first developing and launching its internally developed Home Network Protocol in 2001, launching an integrated Home Automation solution in 2004, and waiting until 2005 to establish a consortium "to standardize home network technology, so that appliances from different manufacturers could communicate with each other." Even with three years of internal development and an investment of more than $12 million, however, the LG smart refrigerator was missing crucial interfaces and ecosystem integration at its launch (LG Electronics, 2009).

Despite an impressive array of high-tech hardware, networking, and media components, this smart product failed to win market adoption. It might be argued that LG simply entered the market too early, ahead of when consumers were ready to adopt the services and integrated products that might have been available in a different ecosystem model. After all, the LG refrigerator and its other smart appliances debuted in the period when dozens of recently booming Internet companies were shutting down every month, including those set up to provide home grocery delivery services and online recipe databases – the partners that LG might have been working with in a more open ecosystem model. Connectivity issues and lack of ecosystem partners were not, however, the main problem with LG's strategy for its smart appliances. Like many smart product makers, LG put too much emphasis on the product and its proprietary features and not enough on building a smart ecosystem that would create smart services to add significant new value to the customer.

In contrast, SmartR planning starts with a model for services and ecosystem partnerships rather than with an isolated product design. Before product launch, the SmartR company would enlist infrastructure and ecosystem partners to provide the bar-coded information to be scanned from incoming food items, targeted coupons, shopping, and diet plan services, as well as links to smart meters and other home appliances. SmartR would publish its API and encourage developers to come up with innovative applications through online publicity, developer forums, and a contest for developing the applications to be included with SmartR's launch. Without these extended and interconnected services, SmartR's capabilities would be limited. In fact, SmartR would seem like just an ordinary refrigerator that happened to have an Internet connection and display screen – very much like a number of smart refrigerators that debuted and subsequently failed to win customers over the past decade.

Having determined that services to facilitate better family nutrition and healthier eating at home will be an important value proposition for SmartR's target market, and that consumers interested in nutrition are also looking for ways to save time and money, the SmartR company would partner with online nutrition, diet, and cooking content providers as well as grocery stores, restaurants, food delivery and coupon vendors. The SmartR product would be designed to learn about its owner's brand preferences, dietary needs, and tastes in multiple ways. Its embedded bar-code scanner would be linked to an Internet-connected database of all common food items (with details about brand, size, average price, nutritional content) making it easy for owners to scan and index every item – whether stored in SmartR itself or quickly scanned before landing on the pantry shelves or the kitchen table. This database would enable a multitude of services in addition to automatically generating shopping lists and providing advice on menu and meal planning. Frugal consumers could opt in to get coupons displayed on SmartR's screen or sent to their mobile phones; as they used certain discount offers, others for related items would be added to their profile. SmartR would order groceries based on the owner's shopping list updates and would remember to apply relevant coupons when it placed the online shopping order; it could also send coupons to the home printer or smartphone if the shopper preferred to redeem coupons in person.

Standardized barcodes that supported the incoming product scan would be interoperable with microwave and cooking appliances to automatically read encoded cooking instructions and set the proper cooking time, with all settings displayed on the touch screen for confirmation by the cook. Links to an online recipe database would make it easy to pull together interesting meal options using just the contents of the kitchen at that moment. SmartR could also generate on-demand recipes and shopping lists for more ambitious meal planning. Ideally, the services linked to SmartR's ecosystem could be customized by individual consumers to create a fine-tuned, personalized service with applications available for every member of the family. Personal diet goals would be coordinated with nutrition and diet partners to create customized diet-conscious meal plans, recipes, and delivery of prepackaged food that matched up with specific diet plans. For even more motivation, the SmartR door display would track the calories consumed at home, and accept input from online food diaries to maintain a cumulative record of calories and weight-loss goals. And for the ultimate in disclosure, support, and inspiration, owners could elect to display the contents and activity of SmartR on their Facebook or other social network pages, sharing their progress and lapses with a network of fellow dieters and SmartR owners.

When owners want a change of pace from home cooking, SmartR uses its Internet connection to provide real-time searches of all popular local restaurants that offer the food selection and price range that the owner selects. SmartR displays a list of recommended nearby restaurants and delivery menus, highlighting options likely to appeal to the tastes and nutritional priorities of its owners, based on their shopping and eating patterns that SmartR has dutifully recorded and data mined.

In addition, SmartR would facilitate energy conservation and personal energy management. A wireless connection to the home's smart electric meter would enable SmartR to serve as a hub for monitoring the electricity usage of other appliances and even heating, ventilation, and air conditioning (HVAC) systems with a smart thermostat connection. SmartR would display messages from the utility company about peak demand periods and electricity pricing fluctuations and automatically cycle consumption based on its owner's preferences and power savings goals. Finally, Smart R would follow privacy and data security best practices. SmartR would not share any information without explicit permission from its owner. All the data that it collected and analyzed would be stored securely on a tamper-resistant embedded module and encrypted before transmission.

SmartR, if it existed, would exemplify the power of a smart ecosystem strategy to position a new smart product as a hub of connected services. The possibilities for adding services based on SmartR's embedded connected intelligence and its ecosystem of partners, developers, and coordinated services are unlimited. The more partners and developers who integrate their products with this ecosystem, the more likely consumers are to rely on the SmartR platform to access and personalize more value-added services, generating strong loyalty and creating new revenue opportunities for SmartR's company.

Developing a smart ecosystem with a large, varied group of partners and rich mix of innovative value-added services and applications is more than a strategy to sell smart refrigerators. A thriving SmartR ecosystem would generate recurring revenues and create a branded service platform that could extend across several industry sectors. SmartR would disrupt the home appliance sector by bringing a game-changing smart refrigerator with integrated services to the market. Reducing the initial price of SmartR to well below that of competing appliances would attract strong consumer interest and help to accelerate consumer adoption, enabling network effects and rewarding ecosystem partners with business from enthusiastic early adopters.

Connected home solutions to date have focused too much on connecting products to each other, or on enabling data exchange from a smart product to a designated vendor or back-end system. The SmartR ecosystem strategy does not require any new home networking solutions. All that is required is a broadband Internet connection to the home and a Wi-Fi access point. The missing pieces of the smart ecosystem are the services, not the network. If appliance makers and other device manufacturers want to succeed with smart products, they have to begin thinking of themselves as service platform providers with recurring revenue opportunities that come from smart ecosystems, not just one-time sellers of boxes with embedded intelligence and network connections.

The fact that the home remains open to new and innovative smart products and integrated smart services is a testimony to a deeply felt, if not explicitly stated, consumer preference for a flexible Federator or Transformer style of connected home ecosystem as compared to a Charismatic Leader or Hegemon control-centered model. The reasons why this counterintuitive ecosystem preference may be working to the benefit of the consumer is discussed in the next section.

## LESS VENDOR CONTROL, MORE SMART PRODUCT OPTIONS

More than a decade after the flurry of announcements from home networking technology and platform providers, no one dominant ecosystem has won the battle for home connectivity. Despite the apparent chaos in connectivity options, the lack of Hegemon control is a good thing for consumers who want to manage the home network environment to suit themselves.

Thanks to the ubiquity of the Internet and cable connections, the US home is well served by local and wide area networks. Consumers have several options for connecting to the Internet at home, including the cable broadband network service from their cable TV provider, high-speed fiber phone lines from landline carriers, and wireless 3G network cards for their laptop computer from a wireless carrier. It is even still possible to use a dial-in modem to connect the computer to a landline phone and establish a connection to one of the remaining stand-alone Internet service providers such as EarthLink or AOL. A Pew Research study estimated that

63 percent of all US adults have a broadband Internet connection at home as of April 2009, and an additional 9 percent used a dial-up modem and phone line to access the Internet (Pew Research Center, 2009).

Home networks have remained open to new connections and interconnected smart devices and services because the Internet itself has become a common and ubiquitous platform and different industry sectors have gained a foothold without being able to exclude competitors or eliminate standardized, open alternatives. Proprietary networks and ecosystems coexist in most homes, and service providers in different sectors have not succeeded in establishing control over networks other than their own. As a result, smart home entertainment devices run over several types of networks while the desktop computers may use another type, and the laptops and wireless devices yet another. If a landline phone was still in use, the original wiring for that may be connected to the telephone pole outside. If the bundled offer from the cable provider has been accepted, the home might have broadband cable connections to the Internet and a digital VoIP phone along with the TV cable box control center. The cable or satellite TV connection will in its turn sprout an assortment of wires connecting televisions, video and DVD players, some type of DVR box, and possibly video game boxes, all of which may be connected to a HD TV.

For multi-computer households, there may be a home network router that enables many networked devices, either wired or wireless, to use a single Internet connection. This may include a firewall and security system that protects the home computers and connected devices from the relentless public Internet attacks along with a connection to a Wi-Fi access point. As home Wi-Fi has become more common, many smart devices have embedded Wi-Fi chips, providing the foundation for multiple wireless networks around the house. Wi-Fi networks make it easy for laptops and many wired and stand-alone entertainment devices to connect with their home base station, with the Internet, and with Wi-Fi-enabled peripherals such as speakers and display panels. Game boxes with Wi-Fi connections can access the Internet directly rather than relying on the TV cable for connections to download new content and enable real-time interactions with multiplayer gaming communities. According to TDG research, by 2012 approximately 190 million households worldwide will use a next-generation game console and 80 percent of these households, or about 148 million, will connect their gaming system to the Internet.

In addition to the multiple networks that are connecting various smart products based in the home, the majority of consumers now carry around their own personal, multifaceted wireless networking device in the form of a cellular phone. If this is a 3G smartphone, or a phone with Wi-Fi and Bluetooth wireless connections, then it may soon take on a new role. The smartphone could become the universal wireless device for connecting and controlling the components of the smart home. Smartphones can serve as remote controls for smart devices and act as a user interface (UI) for managing giving instructions and getting data from devices that do not themselves have separate graphic displays or interfaces. This would reduce costs since the smartphone has its own graphical interface, application hosting platform, and integrated support for wireless broadband and local area communications through multiple network protocols.

Another advantage of adopting the smartphone as smart home controller is that consumers are already using their phones for managing other aspects of their lives and for sharing data and graphics and social connections with friends and family. This makes it easier to personalize the home control functions to the specific needs and priorities of different family members and for everyone to participate in the smart home value proposition. Medical devices with embedded Bluetooth can use the phone's Bluetooth wireless connection to upload data from a personal health device such as a glucose monitor to the phone, using it as the bridge between a local Bluetooth network and the Internet, sending regular readings back to a central online medical data storage service such as Microsoft's

HealthVault or directly to the doctor or to a patient's family members. Other wireless networks such as ZigBee are finding their way into the home as well. ZigBee chips are built into smart meters and home energy monitoring devices as well as medical devices and general home automation systems that manage lighting and security systems.

As we have seen, all of these layers and networking alternatives make it complicated and sometimes almost impossible to connect everything-to-anything in the home network nirvana scenario promised by Sun and other early home networking technology providers. In the absence of any dominant Hegemon network, the home is the consumer's own personal technology platform and the consumer becomes the Federator responsible for making choices about when and how to connect an ever-growing collection of smart products. While media producers, entertainment device makers, broadcast companies, and cable providers engage in a prolonged battle to defend their business, copyright, and technology territory, home consumers can decide whether to set up their own home network using Internet and Wi-Fi connections with point-to-point solutions that are available for free or at a low cost from small vendors, or simply live with separate smart product and home network silos. As anyone who rushed to install the first generation of home Wi-Fi access points can testify, early adopters of smart home networking products will frequently pay a higher price in time as well as cost. On the plus side, however, with no Hegemon dominating the home connectivity ecosystem and imposing interface rules and restrictive terms of network use, consumers are free to try and connect as many new devices as they wish. Each household can also experiment with new types of smart services as they become available.

The smart home provides ample space for emerging industry players to try out their product and services offerings over time. If the first foray into Internet-connected smart refrigerators and a networked kitchen hub doesn't win over mass-market customers there is likely to be a second generation attempt or a completely different smart service and ecosystem model that will eventually take hold and become a must-have for tomorrow's consumers. This openness to new solutions and smart products has made the home an attractive but very challenging market for small technology companies and new entrants. On the one hand, almost any innovative smart product will find a few early adopter customers willing to try it out. On the other hand, very few new companies have a chance to reach the majority of consumer households without making strategic alliances with established brands and demonstrating that their product is based on a widely accepted standard and able to run on one or more existing home networks.

Harkening back to Chapter 6, home energy monitoring and personal energy management are in the smart home spotlight today as smart grids and smart meter installation attract government investment and media attention. Despite the barriers to creating smart energy services discussed in Chapter 6, dozens of start-up companies and a number of well-established technology leaders such as Microsoft and Google are rolling out first-generation products that allow consumers to review the details of their energy consumption at home. Many of these product companies are aiming to partner with utility companies to support a direct connection with the home's smart meter. Whirlpool, GE, and other appliance makers have announced their commitment to producing smart home appliances that can connect wirelessly to the smart meter and respond automatically to pricing and demand signals from the utility company. These energy-aware appliances are, however, still several years in the future. Many of today's smart energy start-ups and their products are likely to be integrated into larger solution providers or out of business before consumer-activated home energy management systems become widespread.

The next section will discuss the reality of smart, connected energy and health products as of 2010 with an analysis of three companies that are building ecosystems to serve today's consumer.

CONSUMER CONTROL AT A PRICE

Control4 promotes itself as the operating system for the home, and promises consumers one central control system for "essential energy management, effortless entertainment, whole-home lighting control and comprehensive security systems" (Control4, 2009). This pitch sounds like the "Whole Home Control" solution that appealed to only a minority of consumers in the CABA Connected Home Study, and in fact whole home control is the essence of the Control4 value proposition. Getting an extensive home entertainment system and a basic lighting and security system to work through a single controller costs more than most consumers are willing to pay – between $5,000 and $10,000 – plus the services of a professional installer since Control4 is not available as a do-it-yourself package. Despite this high price tag, Control4 has had considerable success attracting customers: it announced in August 2009 that it had shipped its one millionth ZigBee-enabled device.

The Control4 solution works by communicating with ZigBee chips that ecosystem partners have added to lighting and security systems and other home devices to connect them to ZigBee mesh networks that interface wirelessly with the company's central control panel and remote controllers. Like other universal remotes, the Control4 system also interfaces with thousands of different proprietary home device drivers, using IP and other protocols as needed to get each brand's product to do its bidding. Since the system is installed by a professional, the consumer doesn't need to worry about the challenges of making these connections work as desired. They just have to be willing and able to pay for the end result.

Recognizing that the higher-priced home automation solutions appeal to a small niche market, Control4 has worked to build an extensive Federator ecosystem that is attractive to market leaders in several industry sectors, including home entertainment device makers and utility companies, and that will appeal to independent application developers. It has developed a separate and considerably cheaper product suite for the home energy management market, combining a smart thermostat with a wireless ZigBee connection and a color screen display panel to manage the home HVAC and display DR and peak pricing information from the utility company. The control panel includes step-by-step instructions for programming the smart thermostat, information about energy usage, and consumer-friendly applications such as weather reports and digital photo displays.

While the energy management controller can be integrated with the Control4 home automation solution, it can also be installed as a single smart energy module as well. If enough utilities decide to recommend or subsidize the Control4 module, it hopes to become the Federator platform of choice for home energy management. However, as a reviewer of the Control4 module points out, the competition for that energy Federator position is fierce and early entrants will risk being overtaken by larger players as the market matures.

> While everybody has been milling about, Control4 has built itself into the closest thing yet to an integrating platform. Its lead is due partly to its technology, partly to its channel strategy, and partly to its partnering programs. But will it be able to maintain its lead, with some of the world's richest tech corporations – Microsoft, Apple, Intel, Cisco, Google – poised to jump into the category?
>
> *(Berst, 2009)*

Whether or not Control4 can survive as a provider of home automation and home energy management solutions, its embrace of standards such as ZigBee to build an advanced technology platform for its ecosystem has given it a lead in an still-emerging market sector.

## CONNECTED HEALTH

The next frontier of smart, connected homes is in hosting and integrating a variety of fitness, health monitoring, and medical devices that collect and report data on the consumer's physical condition and well-being. In the context of a trillion dollar-plus healthcare industry and the goals of the US Government to reduce total healthcare costs, smart products that manage chronic diseases and allow aging residents to maintain independence at home are winning adherents and attracting investors. Connected health and medical monitoring devices are expected to evolve into fully-fledged smart home services in the coming decade to help patients manage conditions such as diabetes and heart problems. DataMonitor projects that the market for home health monitoring and telehealth services in North America and Europe will grow from $3 billion in revenues in 2009 to around $7.7 billion by 2012 (DataMonitor, 2009).

The bad news on smart health monitoring devices is that the connected health field has already sprouted at least as many network interfaces and communication protocols as home entertainment devices. ZigBee has been adopted by many manufacturers as the preferred connectivity and embedded communications protocol. But some devices are using Bluetooth for communications while others are communicating via Wi-Fi, ANT or smart phones. Consumers who want to connect a variety of personal health and fitness devices in their home are still in the position of having to be their own network administrator and general contractor.

There is some good ecosystem partnership news, however. Major corporate players are forming alliances and creating Federator platforms that provide a single point of use for health providers and consumers and are welcoming device makers, application developers, and start-ups to their smart health ecosystems. One such ecosystem is the Continua Health Alliance which has created device and data communication guidelines and which provides interoperability testing and compliance services for its members, an important component of ecosystem development.

Like home energy monitoring vendors, many connected health device makers have not demonstrated that they have a sustainable business model. It is difficult to motivate consumers to pay directly for connected health services, especially since the people who stand to benefit the most may have limited resources because they are too ill or too old to work outside the home. Many vendors undertake to get their smart device approved for private insurance and/or government health program reimbursement so that the cost of the monitoring service will be paid by an insurance provider. Obtaining such approval is a long-term process, often requiring clinical trials and extensive documentation of health outcomes and effectiveness. This work must be done at the expense of the company, usually several years in advance of the device generating any revenue.

CardioNet, one of the early success stories in connected health monitoring, is an example of the challenges of basing a business model on insurance reimbursement. CardioNet offers a sensor-based wireless monitor that patients with heart conditions can wear day and night. It continually logs heart activity and reports the results to a central databank that is available to the patient's doctor. The monitor detects and reports any changes in rhythm or heartbeat that might require immediate medical attention and alerts the patient as well as the doctor when such changes occur. The CardioNet value proposition for patients with known cardiac problems is the peace of mind that comes from knowing that a cardiac event will be detected immediately even if they don't experience any symptoms that prompt them to call their doctor. The 24/7 log of heart activity allows the doctor to determine the effectiveness of medication or other treatments and to track recovery or worsening of a condition over time. The value proposition for insurance companies is that the patient monitor can notify the doctor in time to intervene before the patient needs to be hospitalized, thus saving on the high cost of emergency and in-hospital care. After extensive patient trials and doctor endorsements of CardioNet's connected health product, major insurance companies agreed to reimburse patients who used it.

However, as insurance companies are looking for ways to cut back on their costs, CardioNet is vulnerable to reductions in the level of reimbursement. In summer 2009 one of the major insurers announced that it was reducing its allowable coverage for a CardioNet patient by 33 percent (from $1,333 to $754), putting a sizable hole in the company's revenue forecast and exposing it to similar cutbacks from other insurers (CardioNet, 2009). It's not clear whether individual consumers can or will pay for CardioNet services by themselves. In a few months, CardioNet went from a high-growth company to one that was pressed to maintain even its current level of revenues.

The alternative to depending on insurance reimbursement is for companies to market connected home health products directly to consumers or perhaps to the families of elders living alone in hopes that the health monitoring will allow them to remain independent. Products that are paid for by the consumer directly must have a clear value proposition, an affordable price point, and an easy-to-understand user interface to give them broad appeal. One example is an early entry into smart, connected health, the Vitality GlowCap™. The GlowCap addresses a common problem – forgetting to take daily medicine or deliberately cutting back on prescriptions to avoid side effects or save money.

Vitality founder David Rose would like to convince insurance and pharmaceutical companies to subsidize the cost of his product, and he has marshaled the data about lost revenue and negative health outcomes when patients do not take their prescriptions as directed. Adherence rates for patients with conditions that require years of pill taking are less than 50 percent, and researchers estimate that almost $200 billion is spent on hospitalizations and in treating the avoidable complications that develop when patients stop taking their medicine (Rose, 2009). For the short term, Vitality is bringing the GlowCap directly to consumers by selling it for $99 on Amazon.

It's a simple concept. The GlowCap fits on a standard prescription pill container and contains a small wireless chip, a sound chip, a colored LED light, and a sensor with a timer that logs whenever the cap is taken off the pill bottle. If the cap stays in place beyond the expected time of day, the pill box cap and a plug-in night light that is also connected to the GlowCap wireless network start to change colors and flash. An alert sounds, and the colors become brighter if the consumer doesn't respond to the first reminder. At a predetermined time, the GlowCap is programmed to trigger a phone call to the consumer's home phone or cellphone with a recorded verbal reminder. If all that doesn't result in any pill-taking action, the next alert goes out to the consumer's family or care provider who can determine the best response.

The GlowCap also keeps an ongoing log with the times medications were taken and any days when alerts were needed. This log can be turned into a weekly report that is sent to the family or the care giver and as directed by the consumer can be sent to a central health record repository such as Microsoft's HealthVault or Google Health. The GlowCap is a very small step in creating a network of smart home health services. But it shows how new entrants can create services based on entry-level smart products to establish a presence in the increasingly crowded connected health space.

## CONCLUSION

Most of the elements of a connected smart home are available as separate smart products and networking components. But the three industry sectors that would benefit most from creating an integrated smart ecosystem for home networking, the entertainment, health, and energy sectors, are competing with each other and contending with unresolved business challenges. The entertainment industry grapples with the threat of content theft and resulting loss of revenue. The Hegemon strategy of utility companies prevents them from partnering with other sectors to develop an integrated home management solution. The lack of clear business and revenue models for connected home health is a deterrent to companies in that sector gaining mass-market adoption in the short term.

The widespread availability of home Internet access and the still highly contested battle for dominance of smart home ecosystems has resulted in a proliferation of smart products and networking options for consumers. This multitude of different smart product and vendor options has the disadvantage of complexity – connecting today's home systems is an expensive project. But for every proprietary option, there is an open and standards-based alternative. On balance this is a positive trade-off for the home owner. Instead of being locked into a single vendor-controlled ecosystem, consumers can explore new technologies and service innovations. Individuals can focus on connecting what matters most to them, whether that is smart entertainment or productivity devices, home energy management systems or connected health services. In making customized connections, they can pick from high-priced turnkey solutions such as Control4, or decide to act as Federators themselves to select the best of the standards-based solutions where the main cost is the time and effort it will take to integrate and configure them.

# 8    Connected machines and consumer value

The machines of the world may not be smarter than human beings, at least not yet, but they certainly have us outnumbered. An estimated fifty billion machines are already at work around the globe, compared to the world population of 6.7 billion people in 2009 (M2M Magazine, 2008). Population growth in most developed countries is slowing, but billions of new microcontrollers and machines go into operation every year. Connecting those machines to each other and to a network of wireless modules and sensors is the business of M2M (machine to machine), a well-established industry with enormous global growth potential.

Until quite recently the M2M sector has focused on solutions for the government and for enterprise customers in security-conscious and infrastructure-dependent industries such as energy resource management, manufacturing, transportation, and logistics. With the proliferation of smart products and connected devices in homes, automobiles, and the pockets of more than four billion cell-phone subscribers, some M2M vendors and wireless carriers are now eyeing the consumer market as part of a strategy to rebrand M2M as smart services. This chapter analyzes the evolution of enterprise smart services from a foundation of M2M systems and the challenge of transplanting a technology and service culture designed for industrial customers and machines into the consumer environment. It describes the benefits that enterprise managers expect from M2M implementations and contrasts these benefits with the requirements for creating a consumer-oriented value proposition for connected smart services.

To clarify the differences between consumer smart services and strategies for enterprise smart services that are rooted in M2M systems, it is helpful to start with a brief overview of the characteristics of consumer-oriented smart services as developed in previous chapters. Consumer-oriented smart services:

- extend smart product functionality
- are integrated with a smart product technical platform
- add visible value to the consumer's use of the smart product
- are proactively selected by the consumer user; may be deactivated or turned off as desired by the user
- are available as an optional paid feature or on a free with-product basis.

The next section discusses the M2M market and its evolution.

## CONNECTED MACHINES AND THE EVOLUTION OF M2M

M2M systems designed for the enterprise feature components such as remote cameras, motion detectors, location-tracking GPS chips, and sensors that can monitor changes in temperature, moisture, vibration, and other environmental conditions. These components are connected to each other and to wide area wireless networks, extending an enterprise customer's eyes and ears to remote locations. Connected M2M modules keep a virtual eye on enterprise resources twenty-four hours a day, seven days a week without the need for costly personnel or supervision. Each module logs the data it has been designed to monitor and sends it through the network to a central point for real-time analysis, aggregation, and storage. If something out of the ordinary occurs, or if the equipment being monitored stops operating, the system can be programmed to trigger an alert. Freight companies and fleet operators, for example, can monitor the temperature of cargo that requires refrigeration and respond immediately if a refrigeration unit breaks down. They can also route freight shipments and schedule drivers much more efficiently when all vehicles and cargoes are M2M enabled, getting the cargo to its destination faster or more economically. Once an M2M installation is in place it enables enterprises to manage complex equipment and transportation logistics, to detect intruders, to monitor lights-out operations inside physically secure installations, to improve the efficiency of far-flung operations, or all of these things at once. Managers can immediately send a response team to the precise location of the problem when equipment malfunctions or leaks are detected. Even when everything seems to be operating normally in the field, analysis of the data collected from multiple M2M modules may reveal a weakness in the system or detect early symptoms of a pending breakdown.

Thousands of enterprise customers around the world rely on M2M for monitoring and alerts. But the vast majority of the world's machines remain unconnected and unmonitored. ABI Research predicts that ninety-five million cellular M2M modules will ship in 2013; of these about thirty-four million will be for telematics (vehicle-based embedded wireless for cars and trucks) and thirty-nine million will be for telemetry (ABI considers smart utility meters, Point of Sale [POS] terminals and remote monitoring and control applications to be telemetry) (ABI Research, 2008a). Even accounting for millions more satellite-connected modules, it is clear that the M2M industry still has plenty of room for growth before it reaches its full potential.

The discrepancy between the enormous opportunity in connecting billions of machines and the current rate of market penetration for M2M raises questions about what is holding back adoption. Industry analysts point to a number of challenges, such as the lack of global, cross-industry standards for interoperability, the custom design of many installations to meet customers' specific machine connectivity and monitoring needs, the high cost of initial implementation, and the recurring data communication and service fees. M2M vendors typically focus on a few specific high-value market sectors, such as natural resource exploration, equipment monitoring, or alarm systems, giving rise to hundreds of small vendors and thousands of installations that are difficult to integrate and upgrade. The emphasis on customized solutions makes the task of M2M system design and implementation expensive and time consuming, reducing the potential for economies of scale and beneficial network effects in collecting and analyzing data across many systems.

Alex Brisbourne, the President of Kore Telematics, an M2M wireless service provider since 2002, agrees that the industry has been stuck on the threshold of faster global growth for a long time. As he puts it:

> The gestation cycle of M2M as a global industry would make an elephant look speedy. For the past decade there have been predictions that M2M devices would connect a significant percentage of the world's machines and it hasn't happened yet. So there is always a nagging question about how to get M2M to scale to the next level and why it's taking so long. One answer is that the markets have been quite fragmented in terms of the number of suppliers and the complexity of the value chain. The majority of M2M applications and services are still built from the bottom up – using embedded wireless to solve a particular problem for a small group of clients in a highly specialized way. The result is that we haven't seen too many multi-billion dollar corporations entering the M2M sector as broader solution providers, even though plenty of them have become customers.
>
> (Brisbourne, 2009)

Brisbourne argues that these factors are changing. He believes that M2M is entering a period of rapid expansion and consolidation. Even more importantly, larger players, including wireless carriers, are paying more attention to the M2M sector. Brisbourne predicts that as larger providers start to compete for business, installation costs will decrease and M2M will become cost-effective for smaller enterprise customers. He anticipates major expansion for the industry in the next few years, commenting:

> M2M has just recently achieved some key inflection points that are going to drive faster growth. Just look at the growth patterns

within the cellular and wireless industry and how long it has taken to get to the current penetration rate. After years of analog wireless service availability, there were only about 5 million cellular subscribers in the US in the early 1990s. The prices were high and access to cellular service was not standardized or nationally available on a consistent quality basis. That's been the story of M2M for the past 10 years. When digital cellular took off, the cost came down and the capacity and quality went way up along with the numbers of subscribers. M2M is just entering its 2G period, with the implementation and communication costs coming down and network availability and performance going way up. Cellular carriers are now very interested in delivering M2M services and multinational solution providers are entering the market. The focus is shifting from simply enabling monitoring to delivering smarter applications and more sophisticated data aggregation and analysis that can be built on top of these billions of connected intelligent endpoints.

*(Brisbourne, 2009)*

Lower prices and better-quality wireless coverage will address some of the barriers to M2M growth. Vendors are also hoping that delivering strategic services based on the data that connected machines are collecting and reporting will provide higher revenue margins for them and encourage their enterprise customers to invest more in M2M installations. Vendors are expanding their solutions beyond simply implementing M2M connections to providing data aggregation and data-mining services for enterprise customers. And as Brisbourne predicted, global wireless carriers are rolling out new M2M platforms and data services solutions that feature smart services. Vodafone launched a new M2M service platform in July 2009, highlighting its support for smart metering and connected cars as well as remote monitoring of equipment (Vodafone, 2009). Qualcomm announced the formation of nPhase, a joint venture with Verizon Wireless to provide an end-to-end solution for M2M smart services, which nPhase defines as the business strategy outcomes of wireless monitoring, data collection, and analysis for the enterprise.

Wireless carriers have a strong motivation for emphasizing smart services delivery rather than the basic M2M connectivity and data transmission. Since many existing enterprise customers have adopted M2M solutions to save money and increase operational efficiencies, enterprise managers are intent on keeping down the cost of data communications and recurring charges. As cellular coverage becomes more widespread and competition increases, the revenue margins for providing data services to M2M installations are decreasing rapidly. To avoid being relegated to the role of a commodity network charging low data rates, carriers have begun to participate in the more profitable business intelligence and analytics opportunities that flow from providing M2M services. Adopting a smart services delivery strategy has business and revenue advantages for the M2M vendors too. Services yield continuing revenue streams rather than a one-time installation and implementation fee. A focus on services also allows the M2M vendor to maintain profitability and generate new revenues in the face of downward pressure on the price of the sensors and wireless modules and other hardware components of an M2M installation as competition increases and system components are standardized and commoditized.

The next section discusses how M2M industry participants are defining smart services for the enterprise market and analyzes the strategic value that smart services based on M2M data could provide to enterprise customers.

DEVELOPING ENTERPRISE SMART SERVICES

Corporations and government agencies contract for M2M installations based on their need to remotely monitor specific equipment, transportation logistics or infrastructure. Decision makers within the organization have already identified that need, calculated the costs of monitoring, and determined that the benefits make it a worthwhile investment. Basic M2M service provides value by alerting the manager when equipment breaks down or something else goes wrong in the field. What does it mean from the point of view of an enterprise customer if these basic services become smarter?

The smart services term has been used in a number of different ways to describe a variety of business service offerings. Even M2M service providers who have recently adopted the term apply it to different types of activities. M2M Premier defines smart services in the context of specific benefits that an M2M customer would receive, describing it as the process of networking equipment and monitoring it at a customer's site so that it can be maintained and serviced more effectively (M2M Premier, 2009).

Qualcomm defines smart services more expansively in terms of value-added, post-sales interactions between a manufacturer and the new services that the manufacturer could offer its customers based on data analysis, summarizing on its website as follows:

> At their core, Smart Services are differentiated post-sales product support, enabled by wirelessly capturing and analyzing timely product performance information. Usually, these services are delivered by manufacturers or service providers to the owners/operators of the serviceable equipment or machinery. Smart Services represent a way of doing business where the relationship a manufacturer cultivates with its customers after the initial product sale is just as financially relevant – if not more so – than the sale itself.
>
> *(Qualcomm, 2009)*

A 2005 article by Allmendinger and Lombreglia provides a more strategic definition of smart services and the value that these can provide to enterprise M2M customers. In describing the differentiating characteristics of smart services, they state:

> Smart services are a wholly different animal from the service offerings of the past. To begin with, they are fundamentally

preemptive rather than reactive or even proactive. Preemptive means your actions are based upon hard field intelligence; you launch a preemptive strike to head off an undesirable event when you have real-world evidence that the event is in the offing. Smart services are thus based upon actual evidence that a machine is about to fail, that a customer's supply of consumables is about to be depleted, that a shipment of materials has been delayed, and so on.

For customers, smart services create an entirely new kind of value – the value of removing unpleasant surprises from their lives. Meanwhile, because the field intelligence makes product performance and customer behaviors visible as never before, manufacturers gain unprecedented R&D feedback and insight into customers' needs and can provide even greater ongoing value.Finally, because it is impractical to deploy humans to gather and analyze the real-time field data required, smart services depend on "machine intelligence." In a smart services environment, reliable and blindingly fast microprocessors do what they are very good at doing: digesting billions of data points, talking to one another about the data, controlling one another based upon the state of the data – all in a matter of nanoseconds. Humans cannot do this, nor should they; this incessant stream of business information should be invisible to people. At the same time, all this background activity gives managers and decision makers much more visibility into a business's assets, costs, and liabilities – precisely when they need or want it.

(Allmendinger and Lombreglia, 2005: 131–145)

Allmendinger and Lombreglia present examples of vendors delivering M2M smart services with the potential to make their enterprise M2M customers more productive and profitable. These services are based on a more in-depth analysis of the data that is already being collected by M2M installations, and a readiness to act on that data on behalf of the customer. Examples include "proactive upgrading and servicing" such as providing automatic software and firmware updates, automatic deletion and addition of firmware, or scheduling service calls on a pre-emptive basis when the equipment monitoring data indicate that a service call is justified. Data analysis can be turned into more detailed "reports on usage" about each monitored installation, such as time and duration of the operation of the equipment, the machine cycles performed and possibly a log of exactly what functions were carried out by each machine operator.

The article's conclusion is that to create competitive advantage for the enterprise and its end-user customers, M2M vendors must go beyond just providing a standard M2M implementation and creating alerts and monitoring reports. The strategic value of M2M comes from follow-up analysis of data that has been collected and aggregated over the long term. With longitudinal time-based data mining, it is possible to detect patterns that cannot be seen in day-to-day operations. The enterprise can build a record of the performance of its products and installations under different conditions of use. It can schedule equipment service calls based on analyzing the average lifespan of each component, and can adjust pricing based on data about service requirements and the need for replacement parts over time. Harvesting the insights that such data provide can make internal enterprise operations more cost-effective and efficient.

Some insights can also be applied externally to create high-value services for end customers in the type of post-sales interactions envisioned by Qualcomm's definition of smart services. For example, analysis of data from multiple M2M collection points and sensors embedded in the machines that an equipment manufacturer sells to factories can be used to predict factory equipment problems before a machine shuts down, improving productivity and saving on repair costs for the factory owner. Tracking how a number of factories use the manufacturer's machines over time and noting what types of repairs are needed across different types of customer installations can lead to design improvements that make the machines more reliable and more valuable to the customer.

The authors conclude that well-designed enterprise smart services have the potential to increase customer loyalty and generate additional revenue opportunities for the enterprise, adding to the value of their M2M investment. In addition to the benefits and services that it offers to its enterprise customers, the strategic M2M vendor is in a position to analyze aggregated data from its entire installed base of customers, connected machines, and sensors to gain important competitive insights. If the vendor manages to place itself at the center of data aggregation for a large product ecosystem, it can benefit from similar insights into the products and services of its ecosystem partners and increase its chances of being the first to develop innovative new solutions.

However it is defined, the providers of M2M-based smart services share a common assumption that the typical enterprise customer is not getting the full value from a standard M2M installation and that most M2M customers will welcome further analysis of the data that has been collected over time. There is also consensus that data analysis is the key to providing smart services. The more data that is collected, aggregated, and analyzed the more valuable the resulting insights and services become.

Collecting, aggregating, and analyzing data to improve operations and monitor performance makes perfect sense in an enterprise context. Most enterprises are continually looking for new ways to reduce costs and improve productivity. Enterprise managers may even be collecting and analyzing similar types of data already, reviewing the performance and utilization of equipment, facilities, remote installations, and even the behavior of their employees on the job. A smart services vendor can help those enterprise managers make sense of the mountains of data that are already being collected and can apply more sophisticated data-mining techniques to harvest critical pieces of information and deliver them as strategic, actionable recommendations. In the pre-emptive approach recommended by Allmendinger and Lombreglia, the smart services vendor can also proceed to carry out the recommended activities whenever equipment service, repair, or replacement is required.

Smart services such as optimizing equipment performance and making pre-emptive repairs map very well into typical enterprise goals and objectives. The enterprise is committed to improving productivity and maximizing profitability in order to deliver value to owners, shareholders, and customers – and to survive in challenging and competitive economic periods. M2M implementation and the resulting smart services contribute directly to this overarching objective. To the extent that M2M data collection, reporting, and analysis can be rendered invisible in the sense that it is carried out autonomously without the need for man-in-the-loop decision making, this is a positive benefit for the enterprise. M2M components surface alerts when action is needed; otherwise they remain in the background, avoiding a barrage of routine information that would distract managers. The long-term data are aggregated and stored until the M2M vendor compiles an analysis of the data and draws some inferences about how to improve productivity and performance even further. These insights can be packaged for the enterprise decision makers in the form of strategic recommendations, or in the words of Allmendinger and Lombreglia, they can be delivered "preemptively" to the customer to enhance performance and head off problems before they happen. This can create a positive feedback loop in a well-designed M2M installation, where the customer becomes more productive and more profitable thanks to the constant monitoring and pre-emptive services provided by the M2M vendor. In the pre-emptive model, reports on problems avoided and increased productivity can contribute to the calculation of return on investment (ROI) from optimizing the company's equipment and operations.

Table 8.1 characterizes some of the key features of the enterprise smart services value proposition that M2M vendors should aim to deliver to their customers according to Allmendinger and Lombreglia.

Table 8.1 *High-value features for enterprise M2M smart services*

Invisible and remote
Pre-emptive control
Problem-centric and purpose-built
Sensor and machine data and interactions
Optimize machine performance and utilization
Increase productivity and profitability
Reduce costs

The characteristics and benefits of enterprise smart services listed in Table 8.1 do not fit very well into a consumer value framework. Individual consumers are not necessarily driven by the desire for greater productivity, nor are they committed to optimizing their lives or their possessions on a twenty-four hour per day utilization basis. There are no shareholders demanding quarterly reports about the individual's profitability and performance and no common set of expectations about efficiency, revenue generation, and bottom-line results. To the extent that consumers do have articulated goals and objectives, these tend to fluctuate in response to social context and external events.

Consumers, therefore, will evaluate the value of services related to smart, connected products in ways that are quite different from the value proposition of enterprise M2M customers. These differences are too seldom addressed when vendors decide to transplant their smart service offerings from the enterprise world into the home and consumer market sectors. The next section analyzes the consumer perspective on the value of selected smart products and summarizes the smart service features that are most likely to appeal to consumers.

SMART SERVICES FOR CONSUMERS
In the USA fewer than 5 percent of households currently have any type of solution in place that fits the model of enterprise M2M. Most

of today's consumer-adapted M2M solutions are security and burglar alarms with motion detector sensors. But the potential market for M2M-type solutions is growing fast as more consumers acquire more smart, connected products. Beecham Research projects that over 100 million consumer products with M2M-type embedded wireless and cellular network connections will be shipped within the next five years. It predicts that consumer smart services based on home electronics, smart vehicles, and connected health and safety products will become high-growth areas (Steimel, 2009).

AT&T views the untapped universe of consumer products and devices as the next big growth opportunity for creating smart services. In an interview with the GSM Association, Glenn Lurie, President of AT&T's Emerging Devices, predicted that every consumer product will come with embedded wireless in a few years:

> If you walk into a major US retailer – a Best Buy, Wal-Mart, Radio Shack or Costco – and you look at what they sell, you find that for almost every device on the shelf the value proposition for the customer would be improved if it was connected. It would simply be better in just about every way, offer more applications and more solutions for customers. Then you take that concept even further into the healthcare world or the automobile world; connectivity really expands so many aspects of our life. The reality is there should be very few things that we do in our lives that aren't connected.
>
> *(Chambers and Grant, 2009)*

Consumer smart services would be a particularly profitable new market for wireless carriers if they could be sold as a separate subscription service rather than as an option that is bundled into the product purchase price such as the Kindle. If smart services based on individual products become as popular as mobile web browsing, email, and video, this would create a significant new stream of monthly subscription revenue. And once consumers become accustomed to a subscription-based smart service and integrate it into their lives, it will be extremely "sticky" in terms of customer retention. Smart product vendors and manufacturers are also eyeing the potential for recurring revenues from delivering smart services to consumers.

Adding wireless connections to products, however, does not guarantee that individual consumers will be willing to pay extra for product-related services. As we have seen in earlier chapters and will discuss in more detail in Chapter 9, smart, connected products already send selected data back to manufacturers and vendors, sometimes without the explicit permission of the customer, because this data reporting is designed into the product. Services based on the data collected through product monitoring, even services that have proven valuable for enterprise customers, may be perceived negatively by consumers who are concerned about privacy and intrusion into their personal space.

Smart services for consumers must reflect an understanding of how consumer motivations differ from enterprise calculations regarding metrics, such as return on investment and bottom-line benefits. This may seem obvious but in fact many of the smart services being offered to consumers today employ an M2M value model that does not resonate with the individual consumer. Two data points from consumer adoption of smart products help to illustrate the differences between a consumer value framework and an enterprise approach to evaluating smart services.

Over 98 percent of households in the USA own at least one color television set (US Department of Energy, 2009b). Unlike other electronic appliances, this level of TV ownership is comparable across all household income levels. Clearly, almost all US consumers believe that a color TV is something that adds value to their lives. On the other hand, most US households do not have any type of burglar alarm and fewer than 5 percent have a smart security system with embedded sensors such as motion detectors, cameras, and twenty-four hour remote monitoring (Steimel, 2009). Despite the claim made by security system vendors that homes with burglar alarms are significantly safer than those without alarms, and the availability of entry-level security systems at prices comparable to the cost of a large-screen color TV with a monthly cable service, the majority of consumers do not see a compelling value proposition for installing a connected home security system.

Home security systems with remote monitoring services have a lot in common with enterprise M2M installations. When things are normal and nothing untoward takes place, the system is invisible to the homeowner. In the background it silently makes a log of activities such as the setting of an alarm when the resident leaves the house and its disarming when he or she returns. Advanced systems support separate codes for each resident, allowing the system to create a more detailed report on the household's comings and goings, with information such as the time of day when doors were secured or opened and who entered a password at each entrance. But such reports are not particularly interesting to the typical consumer. The main purpose of the alarm system is to pre-emptively alert a security service or the police when an intruder breaks in. More generally, the day-to-day value of the system is preventing something undesirable from happening rather than explicitly and apparently making something desirable happen.

On the other hand, most consumers watch television daily. TV provides them with explicit and apparent value in the form of varied content and entertainment that is available whenever they want it. TV watching also provides a social context for sharing experiences and opinions with friends and family. Even with hundreds of channels available, highly popular shows provide a reliable topic of conversation among peers and co-workers. Very few TV fans would argue that ownership of a color television set contributes to their personal safety or helps them to optimize their performance, but that is not a barrier to their willingness to spend money on TV-related services such as cable subscriptions and TV peripherals such as DVRs and DVD players.

While the majority of television sets are not yet smart products in that they do not have their own embedded wireless connections, the peripherals that most households attach to their TVs create a smart home entertainment system. A television combined with a subscription service for premium channels, a video recording device, and an Internet-connected game console, a typical configuration in many households, offers the consumer an adaptive mix of connectivity, interaction, control, and choice. Such connected peripherals provide added consumer value. TiVo and other DVR devices enable consumers to delegate control for recording their favorite programs and add flexibility to their entertainment center by making recorded content available for viewing at any time. Internet-connected game consoles support avatars, communities, and multi-gamer play, creating a social network for gamers. These are all features that reinforce the value of the television set and its ecosystem.

Mobile phones are also good indicators of the consumer value hierarchy. Smartphones with mobile Internet access and applications are among today's most popular consumer products. As of June 2009 there were more than 276 million cellphone subscribers in the USA out of the US population of 307 million (CTIA, 2009). These subscribers accounted for more than 1.1 trillion minutes of wireless usage in just the first half of 2009. Global smartphone shipments in 2009 outnumbered shipments of notebook PCs for the first time, with an estimated 180 million units sold worldwide (Wood, 2009).

Consumers place a high value on their mobile phones and they are willing to spend money on related mobile content and services. While mobile phones offer many capabilities that enhance consumers' lives and productivity, from the overall convenience of voice calling and data access everywhere to the emergency 911 service to productivity tools like calendar and expense management applications, these are not the type of applications that dominate consumer mobile spending on smartphones. In October 2009, the five most frequently purchased paid applications for the iPhone were Cartoon Wars, Backbreaker Football, RedLaser, Skee Ball and, in the most popular top slot, Tap Tap Revenge 3 (Ankeny, 2009b).

Table 8.2 *High-value features for consumer smart services*

Visible and local
Delegated control
Social interactions
Optimization of consumer experience
Increase well-being and enjoyment
Adaptive, flexible solutions

When consumers allocate the time and money to certain smart product purchases rather than others, they are displaying value profiles that are fundamentally different from those of the enterprise smart services customer. Table 8.2 summarizes the aspects of smart services that create value from a consumer perspective.

Applying this consumer value framework to smart products and services in the automotive, home, energy, and health sectors provides useful insights into the prospects for consumer smart service adoption. Prospective smart services do not have to match every characteristic on this list to deliver value to consumers. Judging from consumer behavior in the marketplace, it would seem that smart services that score highly on the characteristics in Table 8.2 are more likely to be valued and adopted by consumers than smart services that score highly on the characteristics in Table 8.1.

How do vendors make a smart service visible to consumers? The smart product itself should be a manifestation of the smart services that it enables. As such, product design and performance are an essential foundation for consumer value. Just as a smart product is judged first on its visible characteristics – the design, the user interface, the performance, and the level of interaction that it supports – a smart service will be evaluated on the basis of impacts and benefits that the consumer can see and understand. While enterprise managers can quantitatively evaluate the benefits of smart services on their company's performance without ever seeing the services themselves in operation, consumers typically do not have relevant benchmarks to appreciate the value of an invisible service. Consumers are likely to use their mobile phones and their game consoles or DVR devices every day – the value proposition of those products is visible. If theirs is the only house in the neighborhood with a smart security system, and no burglaries at all are reported nearby, there is no evidence that the security system actually protected them at all.

Smart automotive systems, for example, are designed to highlight the value proposition of their embedded services in day-to-day driving. Driver safety modules provide regular feedback to the driver on routine maneuvers such as changing lanes, avoiding objects entering the driver's blind spots, and maintaining the appropriate distance to cars in front. Hybrid vehicles do a particularly good job of giving real-time visual feedback to the driver, showing the impact that acceleration, speed, and braking have on how much fuel the car is using and how efficiently it is operating. Instead of working invisibly in the background, the hybrid is constantly reminding the driver of what a smart, fuel-efficient vehicle he has purchased. In the process, the vehicle is reinforcing driver behavior that increases fuel efficiency, providing even more value. Smart automotive services are also designed around delegated control, leaving it up to the driver to process the visual or auditory feedback from embedded safety systems and respond appropriately. Connected automotive services, such as updates that inform the driver about traffic congestion and road conditions based on information collected from other vehicles, are a good example of building social data components into smart services. The smart service is obvious, visible, and delivered locally to where it is needed by the consumer.

As discussed in Chapter 6, lack of any visible service component is one of the many problems with the consumer value perception of smart meters. The smart meters themselves do not exhibit any consumer-oriented service or value and in fact are designed to report data to the utility company not to the consumer. In DR pilot projects, consumer participants who had a separate in-house device that provided visual reminders about peak demand pricing periods were more likely to shut off appliances and adjust their air conditioner and were also more positive about the pilot program experience (Brockway, 2008). Several participants in the Chicago pilot program noted that the visual reminder device was a particularly valuable part of the pilot program. Providing visual reminders is also a strong value proposition for smart health devices aimed at patient adherence to taking prescription medications. As we saw in Chapter 7, the GlowCap smart service has a number of behind-the-scenes options that operate invisibly, such as logging the date and time that the pill box cap is removed and sending alerts to family members if the cap is not removed. But its design features also make it highly visible to – and interactive with – the patient.

Smart service implementation does not require building a new connected product from scratch. Companies with innovative ideas for smart services can build their service offerings into existing smart product ecosystems. RedLaser, one of the top 2009 iPhone apps in terms of popularity and revenue, provides a good example of the development of a smart service using an existing smart product and ecosystem partnership to match the value framework of the individual consumer. Using the RedLaser app consumers can point their iPhone camera at any ordinary one-dimensional barcode. The application then transmits the scanned image through the application to get detailed product and pricing information and product reviews from Google Products Search and Amazon.

Occipital, the start-up company that developed RedLaser and several other smart services for the iPhone, has developed an innovative image enhancement technology that provides much better resolution for barcodes that are scanned with a mobile phone camera. Until recently, camera phones could not deliver adequate resolution to make them reliable scanners for standard barcodes. Most mobile barcode-scanning applications are designed for reading 2D barcodes which are not widespread and are unlikely to appear on low-cost consumer items for the foreseeable future.

RedLaser succeeded in transforming the information that is embedded in the barcode into a form that provides visible and obvious consumer value. RedLaser changes a barcode that is readable only by a machine into a display of product, pricing, and review information on the iPhone screen. The smart service is flexible and adaptable to consumer preferences; it can scan barcodes on food labels, consumer electronics, or packaged goods. Users can read the information provided to make an immediate purchase decision, store it for later reference, or send it to a friend. When they are shopping for a relatively expensive item, consumers have the advantage of seeing product reviews, product specifications, and comparative prices to help them make a decision. Many aspects of RedLaser combine to create a compelling, visible, and very tangible customer experience.

Occipital is also working to enrich a broader smart ecosystem for mobile scanning and visual search services. The company's technology for obtaining better resolution for camera phone images is available for licensing and it has released a software development kit for other application developers who are invited on the company's website to participate in creating new services:

> The RedLaser app lets you search for online prices using Google product search. You can also keep a list of products and send them to yourself or a friend (because sometimes you just need to check things out on a bigger screen).
>
> Oh, you wanted to do all that other stuff too? We know! But we're too busy to make every application under the sun, so we've made RedLaser's technology available for use in other applications with the RedLaser SDK.
>
> (RedLaser, 2009)

The RedLaser application provides significant value for the consumer. Once it is downloaded to the iPhone, there is no subscription fee or extra costs. It is available to use wherever the consumer wants. Upgrades and new features are available for free. The one-time cost for a consumer to download the RedLaser application is $1.99. With the price bar for smart services in the consumer market that low, should M2M service providers curb their enthusiasm and stick to developing enterprise smart services? Not if the price paid by the consumer is only part of the value proposition for implementing smart services. The next section discusses why pricing models for consumer smart services need to be different from those in the enterprise and why offering a low-priced or free smart service to consumers can be a good business decision.

## THE BUSINESS VALUE OF CONSUMER SMART SERVICES

One point of agreement among smart service providers in the enterprise and the consumer space is that data analysis provides the foundation for value creation. That value scales rapidly with increases in the number of smart products and M2M monitors that are connected and actively reporting data, and as the data is collected over longer periods of time. One day of M2M data on the performance of a single piece of equipment is unlikely to yield any useful insights about the long-term reliability and service needs of that machine, even if the data is based on detailed monitoring of the individual parts of the machine, its environment, and the efficiency of its operation. One day of similar data on the performance of a few hundred identical machines is much more interesting; if that data can be combined with information about the date that each machine went into operation it has great potential value. Analyzing a year of detailed data about the performance of a thousand machines has an even higher value; this much data will reveal the patterns that are characteristic of a machine or a part that is about to malfunction, the environmental issues that lead to slowdowns, and other correlations. Such analysis may inspire the manufacturer to change the design of parts that fail repeatedly, add new features that will make the machine more efficient for customers, or design a superior product.

M2M data analysis can yield a wide variety of high-value enterprise-oriented insights. The manufacturer may not have the internal expertise to aggregate sufficiently detailed data and to analyze it for the most relevant patterns. Knowing that the M2M data that is already being collected will yield valuable insights motivates the manufacturer to pay a vendor to provide enterprise smart services.

Similarly, the list of product barcodes scanned by a single iPhone subscriber during one day of shopping has no particular value for anyone besides that shopper. Even a list of the items scanned together with the prices and reviews that were retrieved from Google and Amazon would have very limited value for business decisions if it represented only one shopper. However, if the daily barcode scanning activities of thousands of shoppers with an application like RedLaser were available for analysis, that data would be much more interesting, particularly if those scanning activities could be correlated with data about what items were subsequently purchased, at what prices, and in what geographic locations. Analysis of a year's worth of such data would yield a number of high-value data points and insights for retail merchants, consumer product manufacturers, advertising and marketing agencies, and other consumer-oriented companies. There are a number of potential customers for reports and analytics based on this data; and many of them would be likely to pay a considerable service fee to obtain such reports. At this point a critical difference between the enterprise data analysis example and the consumer example emerges. The enterprise customer whose data is being analyzed stands to benefit directly from receiving the results of that analysis. The potential benefits of productivity and optimization will motivate the enterprise manager to pay at least part of the cost of the data analysis. Analyzing consumer data about the details of smart product and service activities, however, does not necessarily provide any benefits for the individual who contributed that data. Consumers, therefore, need additional motivations to pay for smart services. They may even need motivation to share any of their data with service providers.

Individual consumers are using their smart products for personal reasons. They may well object if their preferences and barcode-scanning behavior are being stored and analyzed by third parties unless they have explicitly agreed to share this information and have assurances that data cannot be traced back to them personally. Even if the individual consumer did have an interest in seeing an analysis of millions of instances of barcode scans, it is not likely that he or she would be willing to pay a service fee just to have access to this analysis unless there is some tangible value associated with it.

The experience in the automotive sector, where smart, connected subscription services have been available to consumers for two decades, provides a useful reality check on the prospects for convincing consumers to adopt smart services on a subscription basis and to pay a premium for them, even when those services do deliver considerable value. The OnStar example also illustrates the different value propositions for consumers using smart services and for the manufacturer, in this case GM itself, analyzing the data from smart, connected products.

GM designed OnStar as a smart service that could accomplish two separate and somewhat different goals. As an embedded connected enterprise system, OnStar delivers large amounts of vehicle performance and diagnostic data directly back to GM. This embedded analysis and reporting function is largely invisible to the car owner. GM surfaces a part of the embedded reporting capabilities of OnStar to create an emergency road service for GM car owners who were willing to pay a monthly subscription fee for using the service. If even half of the consumers who purchased an OnStar-enabled vehicle between 2000 and 2008 had opted to continue using OnStar on a subscription basis for the duration of their vehicle ownership, the program would have generated outsized profits for GM. But OnStar's high monthly subscription cost led many GM owners to discontinue the service, even after GM started to offer a free first year after vehicle purchase to demonstrate its value. OnStar had only about 5.5 million active consumer users in 2009, including new owners still in their first free year of use, and the attrition rate of paying subscribers remains high.

Even with a high churn rate of subscribers for the consumer-facing component of OnStar, the enterprise-facing connectivity components generate significant value for GM simply by delivering diagnostic and performance information about millions of vehicles that are still under warranty. Every GM vehicle with embedded OnStar capability continues to report its vehicle diagnostic and performance data back to the car maker. That allows GM to reap the benefits from detailed vehicle performance data reports, to aggregate this data from millions of connected vehicles, and use the aggregated data to improve the design of new vehicles as well as to become more efficient and cost-effective in internal operations such as warranty servicing. These are the typical enterprise value features of M2M services.

As the number of OnStar vehicles providing feedback to the factory increases, so does GM's ability to detect patterns and to deal with any performance problems proactively – both in servicing cars that are already on the road and using the remote diagnostics data to correct problems with cars still being manufactured. Using its database of performance history and the remote diagnostic data generated by OnStar when a car breaks down, GM service providers can determine what level of mechanical expertise will be needed for repairs and even whether or not to send out a tow truck, a new battery, or just the equipment to jump-start the battery.

Embedded connected systems such as OnStar demonstrate the manufacturer's value proposition for detailed product monitoring and performance data analysis across millions of vehicles. Whether or not GM owners pay a subscription fee, the OnStar module provides measurable benefits for the auto maker. iSuppli Automotive Research reports that GM is reaping significant annual savings from the vehicle performance data it collects through all OnStar-enabled vehicles:

> "There is a strategic value in having connectivity to the digital systems in the car," said Phil Magney, vice president of iSuppli Automotive Research at iSuppli. "While the value of this is hard to monetize precisely, automakers have the ability to reduce warranty and service costs through telematics."

For example, remote diagnostics – i.e. the capability to remotely monitor vehicle conditions – has the potential to save automakers millions of dollars over the long term. While it is difficult to estimate the total cost savings from these technologies, GM has stated that OnStar saves it more than $100 million annually through information that flows back to it from OnStar-enabled vehicles.

(iSuppli, 2009b)

Similar telematics programs, however, have failed to survive the resistance of mass-market consumers to subscription-based smart services. A subscription-based smart service called Wingcast that Ford and Qualcomm launched as a joint venture in October 2000 provided a range of driver safety, navigation, and emergency road services similar to OnStar. Ford shut down the Wingcast service in June of 2002, citing consumer resistance to the subscription services model for telematics. In its statement about discontinuing Wingcast, Ford management said the shutdown was prompted by a lack of demand for in-vehicle communications systems in the automotive marketplace. "It has become clear to us in the last couple of years that customer demand for telematics systems is changing, both in terms of what they want and what they are willing to pay for," the Ford spokesman said (Murray, 2005).

Ralf Hug, Vice President of Business Development for Airbiquity, a platform provider for OnStar as well as other telematics services, agrees that the subscription model is a formidable barrier to customer retention for automotive smart services. However, he points out that the difficulty in monetizing a high-cost embedded telematics platform has led to alternative models that can be supported by different revenue options such as advertising and partnerships. According to Hug:

From early 2000s until fairly recently all telematics communications were based on a high-priced embedded module with wireless capability. Most manufacturers limited telematics

to a subset of vehicles, usually the most expensive models
where buyers were more willing to pay a premium. The Ford
Sync business model works for a lower-priced vehicle because
the consumer brings their own smartphone to provide wireless
connectivity and the Sync leverages the Bluetooth radio in the
phone to connect to the services in the car. Ford Sync uses the
Airbiquity gateway to link to services in the cloud and brings
these into the car through the driver's mobile phone. This model
provides access to all kinds of consumer-facing services; GPS-
based directions, location, infotainment, links to iPod or other
consumer provided music players and so on. Ford Sync started
by offering entertainment-type features, but they are adding
more emergency services and vehicle performance reports as the
service evolves.

*(Hug, 2009)*

Hug sees this strategy as a progression away from the walled
garden telematics approach of embedded wireless smart modules
that were completely controlled by the car makers to a more flexible
open architecture that works as a gateway to bring more service providers into the auto maker's smart services ecosystem. But he points
out that without a factory-embedded module that is integrated with
the rest of the vehicle's diagnostic and performance systems, the
auto maker won't get the M2M option of detailed performance monitoring and data reporting back that has proven to be so beneficial to
GM. Nonetheless, the benefits of enabling successful consumer services may outweigh that drawback and provide alternative sources
of driving and vehicle performance data.

The Ford Sync smart services strategy to which Hug refers
takes advantage of the widespread availability of wireless consumer
electronics that are already network connected. Since the consumer
is providing the wireless connectivity to bring applications and infotainment into the vehicle, Ford avoids the overhead cost of a factory-
installed wireless telematics module. It passes this saving on to the

car buyer by providing three years of free Ford Sync use and a minimal monthly subscription fee for the basic Sync service once the free period is over. This allows a mix of business models and provides more flexibility for the car owner who is willing to use his or her personal phone's data plan as the wireless connection for in-car services. The auto maker can partner to provide free and sponsored entertainment and information content, advertising-supported services, integrated voice calling, speakerphone, and voice search. As new content and application partners join the smart service ecosystem, it is easier for Ford to offer them to the installed base of Sync customers, helping to enhance the value proposition of the service and increase consumer loyalty to the brand.

Enriching the smart ecosystem by adding new platforms and services is a good way to create the type of value that attracts consumers to use smart services. Ford Sync has transformed the business model for telematics services by letting its customers integrate their own wireless device into the telematics platform. The more car owners engage with the Sync services, the more Ford learns about the tastes and preferences of their customers and the more appealing the ecosystem becomes for new partners and content providers that generate revenues for Ford. This consumer value-oriented smart service is also a good business value and new growth opportunity for Ford.

Social data and interactivity are also high-value characteristics of consumer smart services. Navigation manufacturer and service provider TomTom has highlighted these value propositions to convince millions of its customers to voluntarily contribute detailed travel logs to a central database of shared driver information. The more data that customers are willing to share, the more valuable this central repository becomes.

Using GPS satellite connection during travel and embedded intelligence, a PND collects detailed information about the vehicle location, average speed, duration of each trip, the time spent driving on specific roads, the total miles travelled, and the starting point and ending point of each trip. Drivers can view some of this information using the trip log feature of the device. Since most PNDs are stand-alone devices with no outgoing wireless connectivity, consumers must manually link the PND to the Internet through their PC in order to download new maps and other content or to upload the trip log data that it collects. To compete with the burgeoning use of GPS-enabled smartphones for turn-by-turn directions, the PND makers are adopting two-way connectivity and data-sharing capabilities to support value-added features and support premium device pricing.

TomTom encourages its customers to upload information about road and construction changes to keep its maps as up to date as possible. Its Map Share program has logged almost a million user-contributed contributions that update street names and other changes in the central navigational database. TomTom also invited its customers to upload their daily driving logs to help create a smarter advisory service that could take into account the impact on travel time of rush hour traffic or local travel patterns at certain times of the day or days of the week. According to TomTom, over seven million customers voluntarily uploaded their detailed day-by-day travel logs, representing more than 1.8 trillion road miles in the USA with time-of-trip information for each route during different periods of time. As a result of consumer-provided data, TomTom can offer smarter navigation and routing services that are based on projected traffic conditions on alternative routes at the specific time that a customer is travelling. It is using this detailed information to provide new services such as IQ Routes (Business Wire, 2009).

What motivates consumers to share such valuable data with a smart product manufacturer? One factor is their satisfaction with the value proposition of the product. If customers see visible value in the products and services provided by companies such as TomTom and Ford and use these services regularly, they are motivated by a social interactive framework that has the potential to generate even more value over time. Trust in the service provider and the smart product increases as it delivers more value.

CONCLUSION

The value of M2M data collection and the M2M model for smart services for the enterprise do not translate directly into consumer markets. As a result, service providers should not expect enterprise pricing models for smart services to appeal to consumers. Companies that want to expand their smart services from the enterprise to consumer markets need to adapt their value proposition and work at articulating value in terms that make sense on an individual consumer level. A company providing smart services for consumers needs to establish a visible value proposition that encourages regular use of the service. The data from usage may be valuable enough to justify offering the service itself at a very low price to consumers who agree to share that data with service providers and with other customers.

Consumer concerns about the security and privacy of their personal data are barriers to the broader adoption of smart services. Smart service providers must be proactive in establishing and articulating policies for protecting consumer privacy. Chapter 9 analyzes the challenges of establishing privacy and security practices in a smart ecosystem and proposes best practices for smart, connected products.

# 9  Smart product privacy issues

There are many e-commerce success stories in the United States, but consumer privacy protection on the Internet is not one of them. Every click on a website triggers a continuous flow of information reporting, from the navigational path that visitors follow, to the topics that they search and the time that they spend viewing various content. All online actions can be logged and then aggregated with data regarding an individual's actions on previous visits. Aggregated data may also include details about registration on one or more websites, subscriptions to online services, purchases, preferences, and other information that has been collected by marketing affiliates and partners. The popularity of online advertising is driving the mining of ever more detailed consumer information in the quest to monetize websites and deliver highly targeted advertisements and offers. In the past decade this quest has created an internet interactive advertising ecosystem that a 2009 study by the Interactive Advertising Bureau (IAB) estimated was responsible for $300 billion in total economic activity in the USA alone (IAB, 2009).

In an environment where tracking and collecting behavioral data is so closely associated with revenue generation, attempts to protect privacy on the Internet through industry self-regulation and voluntary adherence to privacy best practices have been largely unsuccessful. In 1994 the Federal Trade Commission (FTC) filed its first privacy violation case against an Internet company (Swindle, 1999). Fifteen years and hundreds of cases later, the extent to which companies collect online consumer data and track online behavior and the threats that such data collection pose to individual privacy are still contentious issues. In the United States Internet privacy best practices remain voluntary, and online companies, advertisers, privacy advocates, and government agencies continue to disagree about how to balance commercial interests and consumer privacy protection. These ongoing Internet privacy debates illustrate some of the challenges of protecting the privacy of consumers using smart, connected products.

A burst of e-commerce growth and the development of sophisticated web analytics and data-mining technology in the 1990s prompted widespread concern about the impact of the Internet on consumer privacy. In the European Union, these concerns led to privacy protection requirements that all companies collecting online data must follow, under penalty of prosecution. The USA did not follow suit; instead the FTC urged companies to develop internal and industry-wide standards for online data-collection practices that respected consumer privacy preferences. Industry groups such as TRUSTe™ developed and promoted privacy best practices. In compliance with recommended practice, many companies began to display a summary of their tracking, data collection, and information-sharing practices somewhere on their website, under the title of "Privacy Policy." Companies that collected more extensive or more personally identifiable information were asked to take the additional step of requiring a consumer to agree with that collection, or to opt out of it. Most website opt-in processes consisted of the consumer simply clicking on a button to acknowledge having had the opportunity to review and agree with a lengthy statement of practices. This combination of disclosure of a website's tracking and collection practices and a single click signifying agreement by the consumer became the established mode for conveying privacy and data-collection information to online consumers (Cronin, 2000).

A number of studies in the past decade have discussed the limitations of opt-in and opt-out processes and online privacy statements to inform consumers about data-collection practices. Website privacy policies are now ubiquitous but they are seldom read by online visitors. When consumers do take the time to read such privacy policies, they are often confused by the legal style and technical terminology of the documents. It is difficult to extrapolate from privacy statements the basic facts about what specific data a company will be collecting and how they will be used by that company and its partners. Privacy experts have concluded that the more consumers perceive that a privacy statement is complicated and hard to understand the less likely they are to read it, and studies show that consumers are less likely to read the online privacy statements of brands that they have already done business with and trust (Milne and Culnan, 2004: 15–29).

It's not just the lesser-known online companies that use the Internet to amass extensive data about consumer activities, however. Established, trusted brands may also be pushing the boundaries of privacy in their desire to collect more extensive and detailed consumer information through online tracking and data-collection techniques. Sears, one of the oldest brands in US retail sales, launched a market research program on its Kmart and Sears online sites in spring 2007 that demonstrates the limitations of industry norms pertaining to online privacy disclosures and the lengths to which companies will go to track consumer behavior. Sears invited its online visitors to sign up for a program called "My Sears Holdings Corporation (SHC) Community." SHC was described as a market research program that would help the company understand and serve its customers better. Interested consumers were offered a $10 payment to thank them for their participation and were directed to a registration page on the website to enter their name, address, age, and email contact before downloading a software client for their desktop. As is typical of many websites, the registration page included a small window with a privacy statement and user license agreement that could be read by scrolling through it screen by screen, printing it out, or simply clicking a box to indicate agreement with the terms.

Consumers who took the time to scroll through all the pages of the agreement would have seen the following information about what the software would be doing, specifically that the application would "monitor all of the Internet behavior that occurs on the computer on which you install the application, including both your normal web browsing and the activity that you undertake during secure sessions, such as filling a shopping basket, completing an application form or checking your online accounts, which may include personal financial or health information" (Baker, 2009).

In promoting and describing the benefits of SCH Sears did not highlight information to consumers about the SHC Community program's intent to track and report on every interaction with every website, even those involving secure transactions and password protection. The only disclosure of that tracking behavior was the brief description in the middle of a long online statement. This combination of detailed tracking and lack of adequate disclosure led the FTC to file a complaint against Sears stating that: "Respondent's failure to disclose these facts, in light of the representations made, was, and is, a deceptive practice. The acts and practices of respondent as alleged in this complaint constitute unfair or deceptive acts or practices in or affecting commerce in violation of Section 5(a) of the Federal Trade Commission Act" (Federal Trade Commission, 2009).

Discussing the reasons for the action against Sears, which led to a ruling that Sears discontinue the program and assist consumers in removing the SCH application from their computers, FTC Director David Vladeck focused on the inadequacy of the standard privacy disclosure form to convey the true scope of the data collection that Sears had in mind, saying during an interview:

> Sears said it was just trying to get a better handle on the consumer's online shopping experience so they could better serve the consumer. The fundamental message Sears was selling to the consumer was help us help. They were compiling everything that the consumer did on the computer. The disclosure was, don't hold me to this, but I think it was 17 frames on the computer. So it was legalese, it was not comprehensible, and I don't think a reasonable person who read it would've

understood that Sears was going to routinely download everything that that person did on the computer including financial records, health records, passwords. Yes, there were some lines in the disclosure that did, I think, alert consumers to those possibilities. The message is you have to be more transparent about what it is you're doing.

*(New York Times, 2009)*

Some analysts speculated that the FTC intended the complaint against Sears to be a warning to the online sector that companies must be considerably more forthcoming about disclosing the details of any extraordinary data collection that is taking place on a website (Baker, 2009). It's not clear that such warnings can have much impact given the economic incentives for continued tracking of online consumer behavior. In accordance with the reliance on private sector self-regulation, the FTC can only prosecute what it considers to be the most egregious violations of online privacy. The problem with this approach is that what are generally regarded as normal levels of data collection on the Internet keep shifting as the technology for tracking becomes more sophisticated, making both collection and data mining cheaper and more available to a larger number of online companies. Ten years ago the advice that "There is no privacy on the Internet – get over it" was still controversial. Today it's a statement of the obvious.

At a time when smart, connected products are generating unprecedented amounts of data about the daily lives and activities of consumers in their homes, in their cars, on mobile networks, and in every location that they might visit, the lack of progress on Internet privacy protection is not an encouraging harbinger of things to come. Even if smart product vendors have the best of intentions, the technology and infrastructure of smart, connected products and the Internet of Things represent uncharted territory and abundant temptations to test the boundaries of personal privacy. As more smart products are connected to the Internet, new types of reporting and analysis of consumer behavior are enabled.

Often these technologies operate behind the scenes in ways that are invisible and largely unfamiliar to the product owner. This chapter analyzes the range of smart product data collection and the implications for consumer privacy. It reviews privacy best practice recommendations and considers the options for protecting the privacy of smart product owners in ways that will benefit both companies and consumers.

## SMARTER THAN YOU THINK

After fifteen years of disclosure about Internet tracking, consumers are often still unaware of how intensely their online behavior is scrutinized. There is even less awareness about the multiple ways that smartphones and other smart consumer products can, quite literally, monitor every move we make. The fact that some smart product vendors deliberately keep certain monitoring capabilities under wraps makes it even harder to grasp how much data is being collected.

The Pre, Palm's premier smartphone model, is an example of this behind-the-scenes data collection and reporting. The Pre launched in the United States in June 2009 to generally favorable reviews. On its website, Palm promoted the Pre as a highly personalized, intelligent device with so much insight into its owner's life that it can even "think ahead" to provide reminders and relevant information. As Palm described it:

> ... Palm® Pre™ – a phone so in sync with your life it feels like
> it's thinking ahead for you. Pre pulls your different online
> calendars into one view, bringing you the information you want
> without having to search for it. Pre links your contacts from
> different sources, giving you one place to find what you need.
>
> *(Palm, 2009)*

The type of information that Palm was collecting about Pre owners went far beyond what it discussed on its website, however. Just two months after the Pre launch, a mobile application developer and Pre owner noticed that Palm was using the Pre's embedded GPS module to track his location and was automatically reporting that data back to Palm on a daily basis in the form of a detailed location log. The Pre also reported other information, including a catalog of the applications that had been installed and whether or not they had been used that day, and for how long. If an application crashed, the Pre sent Palm a log of the crash conditions. The developer published a note on his blog about this behind-the-scenes reporting to alert other Pre owners (Bohn, 2009).

Within a day of the publication of this information, the Web was populated with headlines such as "Palm Accused of Spying on Pre Owners" and "Is Your Palm Pre Spying on you?" and "Palm Pre Secretly Used GPS to Report User's Location." Some bloggers and blog reader comments expressed surprise and concern at Palm's use of the Pre to track location, comparing it to stalker behavior. Others were blasé, noting that wireless product vendors were tracking everything all the time, making Palm just one of the crowd. Many expressed frustration that Pre owners didn't know the details of what data was being collected and reported by their phones until it was revealed by a technically savvy owner. Although the FTC did not become involved in the Pre case, the problems with Palm's data-collection disclosure were similar to those of Sears in that the Palm disclosure statements and Pre Terms of Use (TOU) did not convey to many of Palm's customers the full extent of GPS location data tracking carried out by their phone, in particular the Pre's daily reporting back to Palm.

Palm did not confirm or deny the Pre GPS tracking practices, nor did it comment on the implications of tracking of subscriber location without explicit consumer notification or opt-in. Instead it issued a press statement that cited the sections in its privacy policy and in the Pre's End-User License Agreement (EULA) that refer generally to its data-collection practices. In a press statement, Palm emphasized that it was following standard industry practice both in its privacy disclosure materials and in its data-collection practices and skirted around the issue of what specific types of data it was actually collecting, stating:

> Our goal has been to follow industry best practices on data collection, use, and encryption. Like most EULAs and privacy policies, though, the terms tend to get pretty detailed about potential scenarios. And because the terms are meant to notify users about all possible variations, we wanted to err on the side of over notifying rather than under notifying users through the terms of use. So there's really nothing here "beyond the norm" for a EULA or privacy policy.
>
> (Bohn, 2009)

Most online and smart product privacy disclosures and EULA documents can be described as following standard industry practice. But that doesn't help consumers to understand what their products are capable of tracking, or what implied permissions they are granting by accepting the TOU. As reflected in online privacy research and statements by the FTC, the gap between standard online disclosure policies and consumer understanding of actual data-collection practices has been a chronic issue in online privacy (Vladeck, 2009). The Pre incident raises another question of particular concern to smart product owners. Is it really true that monitoring and reporting on the product owner's location throughout the day is just "normal business practice" by a smartphone manufacturer? It is very much in Palm's interest to say so, since that helps to shift the focus from Palm's lack of disclosure about Pre tracking activities back to the ongoing debate about what expectations individuals should have about privacy. In this debate, realistic expectations by consumers and the definition of generally available technology are both factors in defining the appropriate boundaries for privacy protection.

Social norms are an important element in setting legal limits to how far someone can go when tracking an individual's behavior and personal activities. The boundary line within which an individual is entitled to a reasonable expectation of privacy has been the walls of a private residence. Law and social norms are in agreement that we have a right to privacy inside our homes. It is socially unacceptable and illegal to prowl around a neighborhood to peer into windows at night and secretly take pictures of what consumers are doing in the kitchen, just as it is unacceptable for a stranger to barge into our living room even if we have neglected to lock our front door. It is also illegal for someone to position heat and motion sensors outside apartment walls to observe movements inside the apartment that provide detailed information about personal activities. In the physical world, market researchers are not allowed to follow us home from a store and make notes about what we do inside the walls of our home, no matter how valuable it would be for that store to know such details.

When activities are challenged in court as invasions of privacy, the court also considers whether the technology being used for observation is generally available as a guide to what might be reasonable. Technology that is typically used for police surveillance, such as heat and motion scanners, does not meet that availability criterion. General availability of a technology and adoption by ordinary consumers may change reasonable expectations about privacy protection. The widespread availability of powerful telephoto lenses has eroded any expectation of individual privacy in outdoor spaces where someone can snap a picture from a great distance without being noticed. If long-distance photography was considered to be an invasion of privacy, tabloid newspapers would be devoid of photos and many paparazzi would be in jail.

After a period of debate about how the inside-the-home privacy boundary applied to the Internet, expectations also shifted. Consensus emerged that the Internet and the Web are public spaces, the equivalent of a shopping mall or a social gathering, even when we are accessing the Internet on a home computer. When consumers view online content or visit a web store and put items in a shopping cart, they should understand that their actions are subject to tracking and analysis. Using a smart product for personal reasons, however, is arguably more of a private act. Are the boundaries of privacy about retreat even further from the walls of our houses to the insides of our pockets?

With over four billion mobile phones in subscriber pockets around the world, mobile phones certainly qualify as widely available technology. Most of these phones do not yet have GPS capabilities that enable detailed location reporting but the popularity of this feature is rising rapidly. An estimated 421 million GPS chipsets were sold in 2008 and 60 percent of these chips went into PDAs and mobile phones; by 2014 analysts estimate that almost a billion of the mobile phones sold annually will have integrated GPS receivers (Meeker, 2009).

With location-aware phones becoming the norm, companies are eager to use location tracking capabilities to implement new services and targeted marketing based on an individual's location. According to ABI Research, mobile location-based services (LBS) revenues will increase from $515 million in 2007 to $13.3 billion by 2013 (ABI Research, 2008b). The popularity of the iPhone and other location-aware smartphone models has whetted the enterprise appetite for ever more detailed consumer location data. As one analyst puts it, "location changes everything:"

> Thanks to the iPhone 3G and, to a lesser extent, Google's
> Android phone, millions of people are now walking around
> with a gizmo in their pocket that not only knows where they
> are but also plugs into the Internet to share that info, merge it
> with online databases, and find out what – and who – is in the
> immediate vicinity ... Simply put, location changes everything.
> This one input – our coordinates – has the potential to change
> all the outputs. Where we shop, who we talk to, what we read,
> what we search for, where we go – they all change once we merge
> location and the Web.
>
> *(Honan, 2009)*

The availability of such detailed information about an individual's location also raises new questions about who can access the individual's data, where it is stored once it has been reported, and what purposes it can be used for without prior consent. Just as with the collection of personal data on the Internet, the EU has developed strict requirements for offering mobile location-based services. The EU e-Privacy Directive stipulates how much information must be provided to consumers in advance of asking them to agree to a value-added location service, places restrictions on sharing of data with third parties, and mandates that any company storing identifiable location data must make adequate provisions for data security. Privacy advocates in the USA have long argued that linking geographic location to logs of individual activities creates inherently sensitive information that should be given special privacy protection. The detailed information about the location of the product owner whenever a smart product is in use, combined with a cumulative record of other aspects of what the owner is doing with the device, such as the information collected by the Palm Pre, creates a very different level of personal profiling compared to the fixed location tracking of Internet browsing behavior. At this point, however, there is no US equivalent to the EU level of privacy protection for location-aware services and the collection of a consumer's location data.

Detailed location tracking is not the only privacy issue presented by smartphones and other connected smart products. The features that make smart products so intelligent are often the same ones that make them capable of amassing significantly more data about more aspects of their owner's life than ever before. In addition, smart products can gather exceptional amounts of personal data without a consumer even being aware of it. The information that a smart, connected product may monitor and report to the product vendor or to third parties includes the following.

- The history of all consumer use of the product – when the buyer first activated it, how often it is turned on and off, what time of day and on what dates it is used, what specific features are used, whether the owner has contacted customer support or replaced or updated any parts or software (and in some cases from what vendor or website those parts or software were purchased), when parts and software are due to be updated.

- The types of software and content the product has on it, exact times and dates that content is accessed or software is run, what other types of information are stored on the product, and how much storage space is still available.
- Whether the owner has customized the product or added any enhancements or special features (and whether these are authorized by the vendor or not).
- In some cases the details of the content by type (photos, text, video, contact lists, spreadsheets, calendars, etc.) and even by file names and size of files (for example, how many entries are in the owner's phone book or contact list or calendar).
- Whenever the consumer uses the available network connections and what type of message or data he or she transmits.
- Whether other products are operating on the same local area network.
- The owner's home location and any other locations where the product is used, including the type of use by location, by extension creating a trail of the owner's movements and habits of use in different locations over time.
- Cumulative behavior patterns over time combining all of these individual bits of data into a profile that may be correlated with other demographic and financial and online information about the product owner and his lifestyle.
- Depending on the availability of embedded sensors, smart products may also measure and report on the ambient conditions that obtain whenever the product is on or in use, for example temperature, barometer pressure, noise level, and shaking.

These factors represent a transformation in kind as well as in quantity of information gathering and behavior tracking. There is every reason to expect that the number of smart products owned by a typical consumer will continue to grow and that such devices will develop more capabilities and more connections. As smart product features expand, consumers have no way to determine the extent of the data collection that is enabled by all of the smart products and connected objects that are part of their daily rounds. The data collection is continuous, automatic, and invisible to the consumer as well as being easily linked to his or her personal identity as the owner of the device.

A number of studies have established that information in a database which has been de-identified by removing personal identifiers such as social security number, name, and address can readily be reidentified through analysis plus comparisons with additional publicly available information about the individuals in the database. Data aggregation and mining have made it easy to pinpoint the personal identity of a consumer based on just a few data points such as zip code, gender, and date of birth (Electronic Privacy Information Center, 2009). Even without such specific markers of identity, researchers are showing that enough data on preferences in movies or products can be mined and linked to other public data sources such as online consumer product or media reviews and blog posts to uniquely identify many individuals (Anderson, 2009). The data collected by smart products is directly linked to the identity of the product owner. This data could be used to link the owner to other previously anonymous data resources, creating an in-depth and ever-expanding personal profile.

In comparison to the detailed but confusing privacy and data-collection disclosures provided by websites, many smart products do not provide any information about the data collected, and others do so only in the product TOU or EULA. Consumers typically have no opportunity to opt out of data collection or to stop the flow of information between the smart product and the vendor.

Not only does the customer have no control over the smart product's embedded communication channels, he or she may not even be aware that any data is being reported back to the vendor; or that the vendor is able to proactively access information and programs running on the device to change settings or to analyze what the consumer is doing with the device at any time. Often this monitoring and data collection is built in to the functionality of the smart product itself so that the owner cannot effectively use the device without the connection that enables vendor reporting and oversight. There is no way to set a "data collection" switch to an OFF position even temporarily.

In addition to concerns about tracking consumer behavior with location-aware products, the detailed and highly personal data collected by smart devices in connected cars and homes, as well as the spread of smart meters and energy usage monitoring devices, raise additional questions about the boundaries of privacy protection. For example, data collected from smart products in the home, such as the energy consumption of appliances and lighting, can reveal a surprising amount of information about consumer activity in a place where personal privacy has historically been protected – the individual residence. Energy consumption patterns can be used to track the number of occupants and guests, as well as how often specific devices are being used. The next sections will discuss how smart products are challenging reasonable expectations for the privacy of driver information, as well as consumer mobile and in-home activities.

## AUTOMOTIVE PRIVACY ISSUES

As discussed in Chapter 4, the embedded intelligence of many automotive systems remains under the control of the vehicle manufacturers and their Hegemon ecosystem. However, a number of embedded modules record data that is of great interest to third parties, including the insurance sector and the legal community. Even though case law has established that much of this data legally belongs to the vehicle owner, there are many situations where it is accessible to third parties including in accident reconstruction, litigation, and insurance claim investigation. Since most automobiles do not have embedded wireless connectivity, accessing data from the car's smart modules such as the (On Board Diagnostic) OBD module is an identifiable separate step, often requiring specialized hardware and software tools. Wireless vehicle connections enable continual and invisible data reporting from the car to a third party. As these wireless connections become more common in cars, questions about the privacy of vehicle and driver behavior information are increasing.

Event Data Recorders (EDR) in automobiles record key data about the vehicle operation in the seconds before and after a collision. The original purpose of EDR recording was auto safety; The EDR is connected to the car's air bag system and is activated whenever a collision, sudden hard stop or other incident activates the air bags. The capabilities of EDRs and the amount of data that they collect vary widely across auto makers and from model to model. GM was one of the first auto makers to adopt the EDR and it is an important component of the GM OnStar system. The National Highway Traffic Safety Administration (NHTSA) ruled in August 2006 that EDR devices on all vehicles manufactured after 2010 for sale in the USA would have to meet minimum requirements for data collection and for crash survivability of the device. Among the required data-collection elements are pre-crash vehicle speed, engine throttle, brake use, measured change in forward velocity, and driver safety belt use. As with OBD modules, as long as the EDR records the required minimum amount of data, auto makers are free to customize it to log many other aspects of vehicle operation for longer periods of time. While it is not mandatory for an auto maker to install an EDR, NHTSA estimates that by 2010 over 85 percent of vehicles will be equipped with one. As one crash reconstruction expert noted: "EDR data is already changing the landscape of traffic accident investigation and, with the EDR compliance date looming, stands to become an indispensible resource" (Gyorke, 2008).

EDR data is frequently used in reconstructing a serious accident to assess liability of one or more of the drivers involved and
to determine if automotive safety systems were operating properly.
EDR can be used in legal proceedings to determine liability for an
accident and to prosecute drivers for reckless driving. If a driver
is involved in a serious accident and the EDR shows that his car
was travelling far above the speed limit and that no braking action
took place before the accident, this information will weigh heavily against him in court; a number of drivers have been convicted
of reckless driving based on EDR evidence. Car owners have also
raised concerns that EDR and OBD recording can be used by auto
makers to invalidate vehicle warranty coverage or to refuse to cover
repairs under warranty because of EDR evidence the car was driven
at excess speeds (Askland, 2007).

Auto makers and law enforcement agencies are not the only
ones with a keen interest in the details of driver behavior and the
causes of automobile accidents. Insurance companies are eager to
collect as much information as possible about how car owners act
behind the wheel and what kinds of driving styles and patterns of
driving are correlated with high accident rates. Total miles driven,
combined with information about the speed of the vehicle, time of
day, and style of driving could all be combined to provide a predict-
ive profile of driving behavior that could be used to determine the
most profitable pricing levels for automotive insurance policies or,
even more drastically, to deny insurance coverage all together. This
is exactly the type of information that Progressive Insurance and
other insurance companies are collecting as part of their discount
rate options for drivers.

Progressive describes its MyRate™ program as an option for
saving on insurance costs. It also discloses to prospective customers
that their rates could increase based on their driving behavior, saying: "When you choose MyRate, what you pay for insurance is based
on when, how much and how you drive. For instance, if you frequently drive after midnight, or you drive aggressively, the price you

pay for auto insurance could be affected when your policy renews" (Progressive, 2009). MyRate customers have a dedicated data-collection module installed in their car for the duration of their participation in the program.

Since MyRate is a voluntary program in which a consumer knowingly agrees to share all relevant driving data with Progressive in hopes of saving money, but with the full understanding that he or she may end up paying a higher rate depending on personal driving habits, it would seem that privacy concerns are minimal. As long as participation in such programs, which are collectively referred to as Pay as You Drive, or PayD insurance, remains voluntary, and the data that are collected remains separate from other personal information, this is a reasonable assumption.

However, Progressive gives itself great latitude in its disclosure statements about the future use and sharing of all the driving behavior data collected by MyRate modules. Given the trends in data aggregation and analysis, it is reasonable to expect that MyRate driving profiles might be combined with other personal data to develop a comprehensive consumer profile. Such data could be shared with life and health insurance companies, or financial and other service providers who partner with Progressive. If that happens, a record of driving after midnight (defined by Progressive as a "high risk" time period) and abrupt braking could become a barrier to life insurance eligibility or could lower a consumer credit rating. Progressive and other insurance companies could use the data collected from MyRate participants to create more detailed driving profiles of high-risk drivers and use those profiles to assess the risk level of new applicants, and flag prospective customers as high risk because of demographic or other similarities to the high-risk MyRate participants. As more states approve PayD insurance rates, consumers who don't opt for PayD may be charged significantly higher rates, in effect creating a privacy tax for drivers who do not want to agree to constant monitoring of their driving behavior. And PayD types of insurance may become mandatory in some states. California is moving towards a requirement that all auto insurance providers adopt a mileage-based system for insurance rates, motivated in part by a Brookings Institute study claiming that PayD models motivate consumers to cut back on driving and therefore reduce traffic congestion, emissions, and air pollution (Bordoff and Noel, 2008).

The data collected by OBD, EDR, and PayD systems may be just the leading edge of a new wave of data collection about consumer driving habits and behavior. Connecting more cars to the Internet through embedded wireless modules, consumer-owned wireless devices such as PNDs and smartphones, and V2I communications will enable real-time tracking and data collection on every aspect of automotive activity. Since automobiles have generally been considered public spaces in terms of reasonable expectations of privacy and protection from tracking of behavior, this data may be widely shared and analyzed for targeted marketing and driver profiling in the not too distant future.

## SMARTPHONE PRIVACY ISSUES

The four billion mobile phone subscribers worldwide already far exceed the number of PC owners. As more smartphones are used for Internet access and Web browsing, the persistent problems of Internet privacy are compounded by consumer use of devices that can track their physical location as well as other activities such as use of the phone's camera and applications. In the era of walled gardens, limited phone capabilities, and strict control of ecosystem partners, wireless carriers provided a single and highly regulated point of contact for the consumer on privacy issues. With the spread of smartphones and third-party mobile applications, the cast of characters who are collecting and analyzing detailed data about mobile subscriber behavior has increased significantly. Whenever the phone is connected, whether through cellular or Wi-Fi channels, applications on the phone may be reporting data about the subscriber back to a third party. Mobile applications that are used to present highly targeted advertising messages to subscribers represent a new set of privacy challenges.

When consumers download mobile applications from different application stores and third-party developers, each application may have quite different models for reporting details about application usage and other device data. Mobile applications may have direct access to information such as subscriber contact lists, email activity, mobile shopping, and use of the camera on the phone. In addition, the wireless carriers and the operators of the mobile application stores may collect personally identifiable data on consumers, for example financial and credit card information. As with other smart product behavior tracking, most of this data collection is not visible to the individual subscriber, nor is there a standard process for disclosure of the tracking and data collection that might be taking place across the entire smartphone ecosystem. As smartphones become more capable and more popular, the development of mobile marketing best practices and privacy guidelines is needed to address concerns about inappropriate data collection in this highly personal context.

Smartphones enable targeted advertising based on real-time information about the individual's location plus a detailed profile of individual preferences and past behavior. The combination of rapid adoption and in-depth profiling capability has made the smartphone a very attractive advertising platform. The FTC has already reviewed mobile privacy practices and personal profiling for mobile marketing and, as it did with Internet privacy issues, the FTC has decided to rely on mobile industry self-regulation for the foreseeable future. Voluntary industry standards are taking shape, and the Mobile Marketing Association (MMA) has developed best practice guidelines for all companies engaged in mobile marketing, noting that: "Current internet marketing and privacy standards do not adequately address the specific challenges faced by marketers when marketing through the mobile channel. Strong mobile industry privacy principles will protect the mobile channel from abuses by unethical marketers, and limit consumer backlash and additional regulatory scrutiny" (Mobile Marketing Association, 2008).

The MMA code provides five guidelines for consumer privacy protection as follows:

- notice of the marketer's identity and the key terms and conditions that govern an interaction
- choice and consent, including an explicit opt-in from the user and a simple and easily discoverable termination process
- customization and constraint, that is the program should reflect national expectations and comply with applicable laws and should be limited to what users have actually requested
- security, defined as reasonable procedures to protect user information from unauthorized use, alteration, disclosure, distribution or access
- enforcement and accountability, which consists of mobile marketers evaluating their practices to certify compliance with the code.

*(Mobile Marketing Association, 2008)*

These are useful principles, but without any means of enforcement beyond self-evaluation by MMA members they are not likely to make much of an impact on the thousands of new marketing and mobile advertising programs that are targeting wireless subscribers.

The rush to open more mobile application stores and to populate existing stores with more diverse applications means that more application developers are participating in wireless ecosystems, and getting their applications onto millions of smartphones. At the same time, Android, BlackBerry, iPhone, and other smart devices are providing more extensive Application Programming Interfaces (APIs) and giving developers the ability to integrate more elements of the phone's hardware and software into their applications. Once installed, applications can carry out many functions independently of the owner, including activating the camera and accessing the pictures stored on the device, accessing and searching the contact list and sending messages, accessing location data, the accelerometer, and other hardware components. The privacy and security issues related to third-party applications on connected smart products have received comparatively little attention relative to targeted mobile marketing, but this area represents a growing risk to consumer privacy.

The BlackBerry, Android, and iPhone operating systems and application stores each have different models for notifying mobile consumers about the types of access that each application has to their data and the access permissions that apps can exercise such as accessing other components on the device.

The BlackBerry gives owners notification about all the function calls and permissions that a downloaded application will exercise. Prior to activating the application on the phone, the consumer can choose whether to "trust once," "trust never," or to "always trust" the application to operate independently. The owner's choice then dictates what level of notification the application will provide with each subsequent use. Opting to always trust the application authorizes it to act independently in carrying out its functions, whatever types of access to other elements of the phone this might require. In practice, an untrusted application will make repeated requests for permissions during every use, making this an annoying and less desirable option than blanket trust from the beginning.

The Google Android smartphone operating system embeds a list of all the permissions and function calls into the application and displays that list to the owner after the application is downloaded onto the handset but before it is fully activated. The owner has two options: enable all the permissions listed or uninstall the application if there is something about the list that makes the owner unwilling to activate the application. Google's application submission process for Android applications requires the developer to disclose all aspects of the application that involve external functions and phone data access. The accuracy of this developer disclosure is enforced by linking the list of application activities that the consumer sees to the actual operation of the application. If a function isn't on the developer's disclosure list, then it doesn't get permission from the Android operating system to perform that function when the application is running.

Apple's App Store and iPhone applications have a "trust me" approach. Apple staff review all applications submitted to the App Store. According to Apple, the approval process includes testing each application for any security issues or illicit data collection. Once the application is approved for distribution on the App Store, it qualifies as "trusted," at least by Apple. The subscriber does not receive any detail about the data being collected or the permissions used when the application is running on the iPhone. Some operations, such as accessing location data, do have to be explicitly approved by the owner, but a great deal of tracking and data collection can take place behind the scenes. The iPhone API gives developers access to information such as the phone's unique identification number, the time and date the application starts and ends and, if the app is, for example, Facebook enabled, it has access to Facebook subscriber information such as gender and date of birth. In addition, each application can track and report all the interactions with the user that take place within the application and can also record and report back on how long and in what locations various application features are used. This is an important advantage for advertising since it is valuable to brands to collect data about how often a marketing message is viewed and whether or not it stimulates additional interaction with the brand such as playing a game, browsing a website, or entering a promotional code.

Given the enormous popularity of the iPhone as an advertising platform, its data-gathering capabilities are not likely to go unused. Advertising is relatively benign compared to more malicious uses of applications to steal personal subscriber data. It's not clear how deeply the Apple review process for App Store submissions checks on potential security and privacy threats under the pressure of reviewing thousands of new applications daily. As we have seen on the Web, many online users never realize that malware is operating on their Internet-connected computers so many iPhones could already be affected by malicious applications. In this case, the control assumed by Apple and the willingness of iPhone owners to trust the Apple brand to protect their security and privacy may create opportunities for exploitation of mobile applications and smartphone subscribers. With over two billion iPhone applications downloaded, the amount of data collection from iPhone subscribers is already enormous and is increasing daily.

## SMART ENERGY AND THE CONNECTED HOME

With the installation of smart meters, the electric utility sector will make a dramatic transition from infrequent collection of limited amounts of consumer energy usage data to a flood of detailed usage data that can be logged at intervals as short as every fifteen minutes. Much of the new data about consumer energy consumption will cross the important privacy protection line that demarcates in-home, private behavior from public behavior that is subject to tracking and analysis without prior permission by the individual. Home energy monitoring devices collect enormous amounts of behavioral information that have significant implications for consumer privacy, but there are few guidelines in place about who owns this information and who may access it for what purposes.

Both traditional and smart meters are typically located outside the home and are owned by the utility company and not the consumer. It is clear from this external location and ownership that all the data collected by a meter also belongs to the utility; in fact meters are designed to be tamper resistant so that consumers cannot get behind the display screen to see, or tamper with, how the meter operates or what specific data it is collecting. These meters may also communicate with smart devices inside the home, such as smart thermostats and energy monitoring devices like the Control4 product discussed in Chapter 7, raising a question about who owns the data that is collected by in-home smart energy products.

In accordance with other court rulings about privacy, the level of privacy protection afforded to consumer energy usage data may depend on whether the data from energy monitoring devices is stored within the home or is automatically communicated to the utility from the smart meter or other connected devices (Mulligan et al., 2009). Smart meters with wireless connectivity send their data directly to the utility company where it may be aggregated and analyzed for various purposes besides billing. Other energy monitoring devices and solutions may piggyback on the smart meter's wireless connectivity to transmit their data or may use in-home wireless ZigBee and Wi-Fi networks. Whatever the communication path used to transmit energy consumption data, very little of that data is stored exclusively on devices inside the home. It is up to the entity collecting and analyzing the data to determine what protection to provide and to disclose those policies to the consumer. An analysis of two popular consumer applications for energy monitoring, Google's PowerMeter and Microsoft Hohm, indicates that individual consumer home energy usage data may not receive any more privacy protection than Internet browsing behavior.

Google PowerMeter and Microsoft Hohm applications are free for consumers to download. Once installed on a PC, both applications link the consumer to a website where home energy consumption data can be compared to regional benchmarks for average households, and current consumption patterns can be used to estimate monthly electricity bills. Both applications depend on connections to partner utility companies or energy monitoring devices to supply detailed energy consumption data. Google PowerMeter receives details about the consumer's energy usage in two ways: through partnerships with utility companies and through consumer installation of a home energy monitoring device with the PowerMeter API and a network connection for reporting the data to Google. According to the PowerMeter Privacy Policy, Google receives the following information from partner utilities about consumers enrolled in PowerMeter:

> Your utility sends Google your electricity consumption information, some historical electricity consumption information, your geographic region (such as your zip code or part of your postal code), and, if available, your electricity cost per kilowatt. This data is encrypted for transmission. The electricity consumption information is sent periodically by your utility to Google and is based on smart meter readings from your household. The frequency of these readings depends on your utility; for example, some utilities will send hourly

readings, others will send fifteen (15) minute interval readings throughout the day. Please ask your utility for details. The electricity consumption information sent by your utility is not broken down by individual rooms, electronics, or appliances in your home. For example, your utility does not tell us whether your electricity consumption comes from the use of a washing machine or a computer server.

*(Google, 2009)*

Google also discloses that it "may use anonymized, aggregate information for internal or administrative purposes" and "may share your anonymous, aggregated electricity consumption information with other PowerMeter users ... or through an API available to developers," giving it considerable latitude about future uses and analysis of all the energy consumption data that it collects. Microsoft's terms of service for similar levels of data collected about registered participants in the Hohm service are even more flexible, stating that "The data collected will be used anonymously and in aggregate for a variety of purposes" (Microsoft, 2009b).

As discussed earlier, "anonymous" data can often be reidentified and associated with a unique individual, especially if information such as a zip code is included. Other studies have demonstrated that mining of hourly data about energy usage may expose information on specific in-home activity and can show if people are using certain appliances or types of electricity even if those appliances are not directly associated with the usage data (Foley, 2008). Law enforcement agencies today frequently obtain a search warrant and use specialized equipment to monitor electricity usage inside a home that is suspected of being a marijuana-growing site, or the location of other illegal activity that involves high electrical consumption. The information being collected about consumer energy consumption by Google, Microsoft, and other companies could be liable to subpoena in the future and could expose consumers to legal prosecution. In criminal hands, the same data could be used to pinpoint times of day when houses are empty or periods when consumers are on vacation. Aside from legal and security risks, detailed energy usage data represents a new trove that can be mined for insights into consumer habits and behavior patterns inside the home, from details about how often consumers use their stoves to what time of day they do their laundry. Consumers may be willing to share this information upon request, but even before they have a chance to consider their personal privacy preferences for reasonable disclosure, the decision about disclosing energy consumption data is out of their control.

The exposure of individual energy usage data is not the only privacy and security risk in the smart, connected home. As more smart products collect information on consumer behavior, many vendors are in a position to develop an in-depth profile of in-home activities. Device-to-device wireless connections such as Wi-Fi make it easy for devices to communicate with each other and share data without the knowledge or explicit permission of the owner. In some cases, such as home health monitoring, medical and health device vendors must follow guidelines for encrypting and securely storing all data. In other cases, such as the data collected from connected computing and entertainment devices, home appliances, and other types of smart products, there are no security or privacy protection guidelines beyond those voluntarily disclosed by service providers.

Consumers have adopted some basic security protection for their home wireless access points, but the easiest default option for connecting new devices to a wireless network is to leave it in the unprotected mode. As more personal smart objects and smart products connect to wireless networks to interact with the vendor or to share information with other network nodes, the chances that a third party is listening and intercepting sensitive data increases. The likelihood of insecure data transmission and eavesdropping is even higher outside the home as the consumer interacts with billions of connected devices and wireless chips that comprise the Internet of Things.

PRIVACY AND SECURITY ON THE INTERNET OF THINGS
Less intelligent and far more numerous than smart products in the home are the billions of wireless chips that have been deployed in M2M systems and enterprise applications. As consumers come into contact with these connected devices and chips, they encounter another set of data security and privacy problems, including those posed by unsecured RFID (radio frequency identification) interactions. RFID chips are commonly used as item tags in inventory control systems and as animal identification tags, as well as in electronic toll collection transponders and in some physical access systems such as the contactless cards that open hotel and office doors.

As RFID is integrated to more consumer items that are directly associated with an individual identity, such as the electronic toll collection tags that are attached to a specific vehicle, this lack of a security architecture means that anyone with a standard RFID reader can access the data stored on the RFID tag or card. The office access card may be cloned for reuse, or the information it contains can be merged with other individual data to identify a specific person.

Basic RFID chips and cards hold a very limited amount of information, usually an inventory control number, a product serial number, or a unique user ID number. Typically, there is little or no intrinsic value in the limited data stored on the chip and consequently little attention has been devoted to securing that data. Most RFID systems are designed to be as cost-effective as possible, and the designers assume that the short wireless range of the chip and the limited value of the actual data provide enough of a deterrent to widespread eavesdropping without layers of encryption and security.

As a result, most RFID tags will simply broadcast their data in response to a signal from any compatible wireless reader that is within range and scanning for a response. Many contactless cards and toll transponders will also identify themselves to any reader that requests information on the appropriate radio frequency. In essence, the unlicensed wireless spectrum is the ultimate party line where any RFID chip will respond to a compatible signal. In a secure system, once the reader selects a particular chip or card to scan, the communication exchange opens with some form of cryptographically protected mutual authentication that further identifies the card to the reader and the reader to the card and may in fact establish an encryption key to protect the data flowing between the reader and the card. There are, however, many contactless card and RFID systems that simply send the card identification number and other data directly to the reader without any authentication. There are also systems where once a card is chosen by a reader, the card continuously and openly broadcasts its stored data until it is told to stop by the reader. This promiscuously broadcast number can be associated with an individual, for example if it is an employee identification card, and used to track people as well as objects.

While each item of information inside an RFID tag or contactless card may be innocuous in itself, when multiple wireless data chips are connected to an individual all it takes is one simple personal identifier, such as a facial image or a registration badge with name rather than random number, to turn the entire set of RFID transmissions into something highly personal, identifiable, and vulnerable to copying by simple scanning equipment. As consumers acquire more items with RFID tags and chips, they present a target for illicit scanning and linking of a variety of data. This vulnerability was highlighted by an RFID-sniffing experiment at the DEF CON conference in 2009. A group at the conference, concerned about the lack of RFID security, set up a table outside a meeting room with a camera silently and automatically photographing attendees coordinated with a standard RFID reader scanning for any free-read chips the attendees were carrying. They succeeded in capturing enough data to identify attendees and clone their access badge and other contactless card data. One of the organizers of the experiment emphasized that anyone could get access to RFID data using a portable reader kit that costs less than $50. A malicious hacker could scan cards and chips in the pockets of individuals at hotels, shopping malls, and office buildings, or the toll payment transponders on the dashboards of automobiles, identify the individuals, and then clone their cards (Zetter, 2009). Some states have regulations restricting the unauthorized scanning or copying of RFID chips, but these regulations would do little to deter ill-intentioned individuals from accessing the data.

## RECOMMENDED BEST PRACTICES FOR CONSUMER SMART PRODUCT PRIVACY

As the number of smart products and wireless chips capable of tracking individual activities continues to increase, protecting the privacy of personal information becomes ever more challenging. What can be done to safeguard consumer privacy short of new government regulations and privacy requirements for smart, connected products?

Privacy best practices can follow two paths. The most familiar and most frequently travelled one is to abide by selected data-collection guidelines and consumer privacy requirements. The second and more strategic approach is to become a recognized leader in respecting and promoting consumer privacy as an important feature of the smart product value proposition. This section recommends best practices for smart product vendors and smart service providers interested in adopting either path.

The following guidelines for smart product privacy protection are drawn from common elements in the FTC publications on consumer privacy and EU Privacy Directive along with recommendations on industry guidelines for privacy self-regulation. Essential components of a smart privacy program include the following.

### Transparency and disclosure

Transparent disclosure should start before a consumer purchases a smart product.

- Vendors will include in advertising and marketing materials, and on product packaging, a statement summarizing the data collection and reporting capabilities of the product along with the address of an online source where the consumer can find more detailed information about how any data collected by the product will be protected and if it will be shared with third parties.

- The online disclosure about product data collection, protection, use, and access will be written in clear understandable language.
- Vendors will seek consumer permission to share data with any parties not named in the online disclosure or to add any data-collection capabilities not disclosed in the original product description.
- Explicit permission and consumer agreement will be obtained for collection of any location-aware information, or any information that is generally considered sensitive, such as financial records, health data, or political and religious affiliations.

## Data security and deletion

- Vendors will protect all data that is going to be stored for the long term as if it contains personally identifiable information even if the data has been anonymized as effectively as possible, because such data may be reidentified.
- All data with sensitive information or identification of the individual consumer that is stored for service delivery, warranty or other previously disclosed purposes will be deleted after a stated period of time and will not be used for data-mining purposes.

## Consumer control options

- Connectivity and data reporting will be controllable by the product owner. Products will include a "data off switch."
- Vendors will not design their products to be dependent on reporting back to home base.
- Consumers will be provided with an easy-to-use template to set their own privacy and data-sharing parameters and send these directly to the vendor. Vendors will confirm receipt and abide by these privacy preferences until notified of a change by the customer.

## SMART ECOSYSTEM PRIVACY AND SECURITY REQUIREMENTS

From the perspective of the smart product ecosystem partners, sharing data is another avenue for providing more personalized services. But in the absence of clear-cut guidelines about what information is to be collected, how and where it is stored, and who gets to access and analyze it for what purposes, it is difficult to believe that there will not be privacy and security problems, especially in light of our experience with Internet data collection.

The ecosystem leader will develop consensus around privacy best practices for smart product and smart sensor network data collection and will establish privacy protection requirements as a precondition for participation in the ecosystem. Requirements should include provisions that all partners must demonstrate their compliance with the ecosystem data collection, storage, and protection norms before collecting any personal customer data. The ecosystem leader will enforce this compliance and will exclude any partners who violate the privacy norms from future ecosystem participation.

### Proactive communication

To go beyond these recommended practices and establish leadership in smart product privacy, vendors must find innovative ways to enhance privacy disclosure and add visible value to their smart products through proactive communication and interaction with customers. One approach is to make the process of data collection more valuable by making it highly visible and interactive with the individual consumer and with networks of customers as appropriate. Amazon and Netflix have demonstrated the value of analyzing online customer data to provide personalized recommendations that improve customer perceptions of service and value. The opportunities to improve service are even greater in the smart product environment.

Limiting the embedded smart product connectivity to a one-way channel and an often invisible process of reporting and collecting what the smart product observes is squandering at least half of the value proposition of a smart connection. Innovative vendors will design connected products with two-way communication channels between the customer and the product vendor. Making the data-collection channel visible and allowing the customer to see what information is transmitted at any time clearly improves privacy disclosure. Equally important, it encourages customer dialog and feedback to the manufacturer, making it easy to ask questions, make comments, and voice complaints and frustrations about product performance while using the device, making the data-collection channel a much more valuable consumer feature.

Given the number of companies that have embraced social marketing and how many companies are paying their employees to write blogs and respond quickly to consumer tweets, it's remarkable that so few are leveraging the connected smart product that their best customers are using every day to encourage direct customer dialog and feedback. There's no button on most smart products to send feedback directly to the vendor, to suggest an improvement or to register a complaint. Smart product connections are an enormous customer communications opportunity waiting to be harvested by innovative vendors.

Vendors could receive even more feedback and data about customer preferences by enabling free social communication among product owners. Rather than sending customers to the Internet to ask their questions and connect with a group of other customers out of view of the vendor, social communication enabled by the smart product would put the vendor in the center of the discussion without violating any privacy best practices.

## CONCLUSION

Tracking consumer behavior is part of life on the Internet, on smartphones, and on other smart, connected products. No matter what new Internet privacy regulations or guidelines are adopted in the future, they are not going to turn back the clock on the practice of personal data collection, data aggregation, and data mining by vendors. Given that hundreds of millions of individuals have voluntarily opted to publish even more revealing personal information about themselves on blogs and social networks, it could be argued that reasonable expectations for personal privacy have shifted towards self-disclosure and that the data collected by our smart products is just another step in the inevitable decline of privacy as we have traditionally defined it.

To obtain the greatest value from in-depth consumer information without compromising consumer trust, companies would be well served by adopting and disclosing consistent and verifiable guidelines for the permissible use, duration of storage, and data security protection measures taken to protect all of the consumer information that is collected by their products. They should also disclose the privacy practices of their ecosystem partners who develop applications and peripherals for those products.

It is not necessary to establish a one-size-fits-all set of regulations for what types of data can be collected or what forms of consent are required before each step. The most important step that companies can take is to make any regular data collection and smart product monitoring activity highly visible to the consumer. Visibility combined with a simple "data off switch" option for permanent or temporary opting out of the collection and monitoring process would create the foundation for informed consumer decisions about their use of smart product capabilities.

Many consumers would opt to accept the vendor's stated data-collection processes in exchange for the value-added services, improved customer support, and personalization that such data collection enables. Those that opt out might do so only temporarily; having control and choice about data sharing would make buyers more comfortable with using the smart product in a variety of ways. And those who opted out entirely might never have purchased the product if that option were not available to protect their sense of privacy. Accommodating this spectrum of privacy options will broaden the base of customers for a particular device. Leveraging the communications capabilities of smart products in ways that allow owners to talk directly to the vendors will create a new form of value and encourage even more data sharing.

# 10 Strategies for managing smart products and services

Smart products are here to stay. Some will enable ground-breaking business innovations, catapulting the companies that launch them into leadership in enormously profitable new market sectors. Others will disrupt traditional market leaders and open industry sectors to a rush of new entrants. A significant number will fail to attract any consumer following and fade from view. Whether or not a company aspires to create a transformational and market-leading new consumer solution, wants to avoid being blindsided by unexpected competition, or needs to understand the impact of embedded product intelligence for its industry and technology roadmap, smart products will play a prominent role in their future strategic options.

There is no guaranteed formula for smart product success, but there are a lot of predictable pitfalls that managers can avoid through a better understanding of the dynamics of smart product ecosystems and embedded control. This chapter distills the lessons from smart product initiatives across industries to offer recommendations for analyzing the impact of smart products on industry sectors, for selecting a smart ecosystem model, and for developing corporate strategies that deliver visible customer value through smart product platforms and services. It concludes that smart products will succeed to the extent that they make the value of smart services enabled by a smart product platform highly visible and accessible to the consumer. Insights about smart product strategy highlighted in previous chapters are summarized as follows.

- To create competitive advantage, companies launching smart products need a strategy to displace the products that are already in use. One such strategy is to disrupt incumbent market leaders and prevalent industry business models. An alternative strategy is to work with incumbent market leaders to add more visible value to smart products and services.

- The strategic value of a smart product ecosystem increases in proportion to the number of ecosystem partners providing value through content and services designed for the product platform.
- Consumer perceptions of smart product value and loyalty to smart products and services increase in proportion to the amount of visible value that the product delivers to its owner.
- Transparency, dialog, delegated control, trust and permission-based personalization are essential components of visible consumer value.

## ANALYZING INDUSTRY SECTORS

Consumer-facing industry sectors have evolved primarily within Hegemon-centric or Federator-centric models. Regulated industries in areas related to national infrastructure or public health and safety, such as the telecommunications and energy sectors, have long been dominated by Hegemon controls and corporate structures. Industries that are capital-intensive and inherently national or international in scope, such as the automotive and broadcast sectors, also tend to be Hegemon-centric. The Federator model is predominant in industry sectors with many small, local players such as construction, retail, and hospitality. Federators also thrive in sectors with many competing standards and platforms and relatively little national regulation such as home products, electronics, computer hardware and software, publishing, and media. Health and medical sectors tend toward either Hegemon or Federator models depending on whether the products and services are primarily medical in nature and subject to regulatory approval or fall into the category of unregulated personal health and fitness.

As discussed in previous chapters, smart products from new entrants are already impacting industry leaders across many sectors. The penetration rate and disruptive impact of smart products and services differs considerably from industry to industry, however. Two important factors affecting penetration rate and disruptive impact

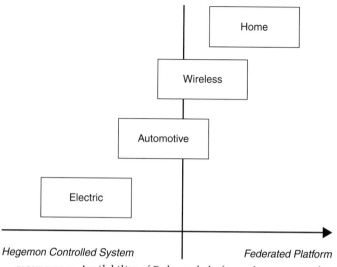

FIGURE 10.1: Availability of Federated platforms for smart products

are the predominant ecosystem model of current market leaders and the availability of connected smart product platforms and their associated smart services offerings. Figure 10.1 positions four industries discussed in this book in terms of the level of control exercised by industry-dominant Hegemons and the availability of Federated smart product platforms to new industry entrants.

In the automotive sector, the dominant Hegemon ecosystem model has accommodated and absorbed an impressive amount of embedded product intelligence in an incremental fashion. In comparison to the average passenger vehicle produced in 1999, today's cars are smarter, safer, and easier to drive. But consumer attitudes and car-buying trends reflect the limitations of incremental innovation in an ecosystem that is dominated by Hegemons and starved for smart service innovation. Regular improvements in automotive performance and safety do not necessarily increase car-owner loyalty, nor do they typically enable smarter platforms to connect the car to innovative consumer services in the broader landscape of smart communications, convenience, and entertainment.

Advanced safety systems and automotive intelligence have progressed almost as far as they can go without crossing the line from delegated control to pre-emptive control and autonomous driving. The investment in advanced safety system development by the auto makers has added to the perceived value of new car models, especially among older and more conservative drivers. It has not been enough, however, to reverse the widespread lack of interest in incrementally improved automobiles among younger consumers. That lack of excitement about automotive innovation has contributed to the current decline in auto sales. Even more troubling to the major auto makers are data that suggest young driver indifference to traditional cars may continue even after the economy recovers. In the USA between 2005 and 2008 sales of new vehicles declined over 30 percent among twenty to thirty-four year-olds, the largest decline of any age group, with analysts noting a parallel decline among young adults in car-related hobbies and leisure activities (Abelson, 2009: A11).

A notable exception to this lack of interest in today's cars is the affinity of eco-conscious young adults for hybrid and electric plug-in vehicles. Toyota's Prius and other hybrid vehicles that command a premium price for their smart technology may not provide any economic benefit for car buyers since the payback period in terms of fuel efficiency is typically longer than the average life of the vehicle. But hybrid cars offer more visible value by interacting with drivers to engage consumer interest in a fuel-efficient driving style. The Prius dashboard prominently displays the transitions from gasoline to electric power along with MPG estimates to create a visible feedback loop between the driver and the engine. Many Prius owners enjoy that feedback loop and use it to keep track of their best mileage results.

Consumer interest in fuel-efficiency optimization is good news for plug-in electric vehicles which will require even more driver engagement to manage power consumption. If electric vehicle adoption does in fact take off, it is likely to include a shift in business models for car ownership; for example consumers may pay for an electric vehicle as a smart service rather than as a one-time purchase. Implementation of such services may involve a version of the Better Place business model in which the car buyer leases the battery and pays monthly for charging, or it may involve leasing auxiliary smart power storage units that allow drivers to manage battery recharging needs as cost-effectively as possible. These shifts may transform the plug-in automobile into a connected platform for smart services of all types, not just those involving the battery and recharging. Any dramatic business model transformation in the automotive sector could be the final blow to the dominance of traditional Hegemon-oriented auto makers that cannot adjust quickly enough to retain control of the industry.

As past failures of new entrants in the wireless and automotive industries have shown, it is extraordinarily difficult to penetrate and disrupt industry sectors that are dominated by Hegemon models. When challenged, the Hegemon leaders are in a strong position to hold out against new entrants and they may invoke the protection of public sector regulators and public safety officials to buttress their resistance to change. Entering such a sector with a smart product platform aimed at disruption is a high-risk strategy. As Apple has demonstrated in the wireless sector, if the new entrant succeeds it will reap extraordinarily high rewards and potentially disrupt the entire industry. The current generation of smartphones and mobile devices with alternative connectivity options has changed the dynamic in the wireless industry and opened the gates to more smart products that can deliver services independently from the Hegemon ecosystems of the wireless carriers.

It has taken more than a decade for smart wireless platforms to gain traction and wireless carriers still retain considerable power and control in the industry. But the lack of innovative services beyond voice calling plans has long been the carrier's weakest point. In the future carrier control will be shared with device makers, unlicensed network services providers, and myriads of application developers as more Federators, Transformers, and Charismatic Leaders emerge. Now that mobile content and application distribution is more open there is no stopping the flood of new ideas for powerful and consumer-friendly smartphone and mobile device platforms.

On the other hand, smart products and advanced technology are not enough to displace Hegemon control of an industry that is heavily regulated. The energy and utilities sector has yet to be disrupted by the impact of smart grid and smart meters and customer installation of renewable energy resources. The grid infrastructure remains firmly under the control of the utilities, backed by regulators and government oversight. The rapid technical innovation and the development of smart products for home energy management, renewable energy generation, and power storage are driven by technology companies and by ecosystem partners of the utilities who are not ready or willing to disrupt the control of current industry leaders. At present, technology leaders positioned to play a disruptive role, such as Google, Microsoft, Cisco, and IBM are joining the utility company's ecosystems, rather than trying to disrupt them. It will take decisive action at the national level to open the US utility markets to broader consumer participation in net generation. Disruptive alternatives to entrenched business models can only create a foundation to make transformation feasible in this sector at a later date. Unlike the state of the telecommunications industry today, there is no alternative smart product platform to challenge the grip of utility companies on the power grid, nor is there an equivalent to the unlicensed wireless spectrum to allow companies to experiment with innovative technologies that will underpin future smart products and services.

A Charismatic Leader or Transformer may emerge in the US energy management sector in the next few years, but that development is more likely to take place in a country where regulations on electricity storage and grid access can be adjusted to accommodate new business models more easily. In the interim, business models based on disrupting the utilities are not likely to succeed. One reason is that the Hegemon utilities will not hesitate to assert, as the landline and wireless network operators did for decades, that connecting any components to the network, in this case the electricity grid, that are not strictly controlled by the Hegemon, will be dangerous and must be prohibited under force of law. That claim turned out to be exaggerated in the case of the network operators – the Carterfone and countless other devices were connected to landlines and wireless networks without damaging the network infrastructure. The actual threat was to the business model of the network operators, rather than to public safety.

In the case of electricity, however, there may be a very real public safety risk of moving too quickly to adopt new technologies that bring power into homes and office buildings separately from the traditional grid systems. There are no early adopters willing to be the first customer to be electrocuted by a smart product. Disruption may be hastened by consumer adoption of electric vehicles and advances in smarter and more affordable residential power storage technologies, or the emergence of microgrid systems that create alternative, unregulated sources of power. These technologies may provide an impetus for innovation in creating new services for selected consumers and enterprise customers who are willing to work through the version 1.0 inconvenience of innovative solutions for energy delivery, storage, and smart management in a residential and small business context. Notwithstanding the impact of new technology, an incremental pace of smart product development in the energy sector is likely to be the norm for some time to come.

The connected home appears to be the sector most open to transformation by smart products, and the easiest sector to enter because there is not yet a dominant player delivering a cluster of integrated smart home services. As many smart product companies have learned to their sorrow, however, it is very hard to move beyond a small group of high-tech gadget lovers to reach the mass-market adoption for residential connected products. Consumers respond negatively to the prospect of adding more layers of technology because their homes are already filled with incompatible smart devices. The plethora of Federator-style technology platforms for the connected home has led to a fragmented market.

A better strategy for the home market, and indeed for any industry sector, is to focus on developing a smart product platform that will support a cluster of high-value smart services. The SmartR scenario discussed in Chapter 7 provides an example of using a services-based business model to bring a smarter version of a widely adopted home appliance or device into the market at a disruptively low price point. Instead of simply charging a one-time purchase price for the smart product, smart product platforms with high-value services offer consumers more options to acquire the product. Developing smart products as platforms for revenue-generating smart consumer services opens up many disruptive and high-growth opportunities.

Small companies can also enter Federator-centric industries by joining one or more Federator ecosystems and adopting the Federator platform to deliver smart services to the target market. For example, the smartphone as a platform provides an entry point to the application ecosystems of established market leaders. Smartphones enable service-oriented applications that can be sold directly to consumers through a mobile application store. This approach allows companies to enter new market sectors quickly and at a lower cost of customer acquisition. It also gives the company credibility and experience with delivering smart services, and paves the way for a partnership with one or more ecosystem leaders.

In the long term, smart products and services will have the most disruptive impact in industry sectors and markets that have been dominated by Hegemon ecosystem models because these markets are slow moving and starved for innovation. For the most part, smart products and services will come from outside the Hegemon walls and will meet with fierce resistance by incumbent market leaders. Once the process of industry disruption gains momentum, however, the more innovative Hegemon leaders may reinvent themselves to create their own Federator and Charismatic Leader ecosystems.

### SELECTING A SMART ECOSYSTEM MODEL

As discussed in Chapter 2, each ecosystem strategy has characteristics that provide advantages and disadvantages to companies launching a smart product and to the partners who join the ecosystem. A company must develop goals for a smart ecosystem in the context of its current market position, priorities for smart product development, and the resources it can devote to supporting an ecosystem model. Since ecosystem partners add value by providing integrated functionality and services for the smart product platform, it is also important to consider which ecosystem model will be most likely to attract high-quality partners. If it is in the company's interests to disrupt current market leaders in a target industry sector, and to attract a large number of partners in a short space of time, then the Transformer and Charismatic Leader models are a good match.

The Charismatic Leader model is attractive for its high impact and rapid growth potential. Strong brands value the ability to keep control of many aspects of the ecosystem to protect the positioning and value of their brand in existing markets. Success as a Charismatic Leader is enhanced if sizeable groups of customers have already adopted a company's products and are highly motivated to adopt industry-crossover smart products from that company because of their loyalty to the brand. Leading-edge technology and services for new product platforms are implicit in the Charismatic Leader model, as is a disruptive business strategy. It is not enough for a brand to simply put its logo on a smart product, or to extend its services into another industry through traditional strategic alliances. As the Apple iPhone and Amazon Kindle ecosystem strategies illustrate, it is also essential for a Charismatic Leader to have enough influence with key ecosystem partners to obtain concessions that will help to disrupt incumbent industry leaders and spur customer adoption of the Charismatic Leader's smart product and services. In the case of

Apple, the key partners for iPhone launch included at least one major wireless carrier in the USA and thousands of application developers. In the case of the Kindle, the key partners are the publishers who have agreed to sell e-books through Amazon at a price point that is attractive to consumers.

While the rewards for success in this ecosystem are very attractive, relatively few companies are positioned to become Charismatic Leaders. Large companies that have well-established brands may be hampered by current industry partnerships with Hegemon leaders that prevent them from effectively developing disruptive business models, as illustrated by the difficulties that Nokia has faced in establishing itself as a Charismatic Leader in the wireless sector. Brands that try to straddle the line between Hegemon and Charismatic Leader, exercising strong ecosystem control without offering partners innovative and profitable business opportunities, are limiting their appeal to both partners and to customers. Smaller companies rarely have the ability to attract and influence the number of ecosystem partners required for rapid growth of services and functionality of their product platform. However, it is not necessary to lead an ecosystem to share in the benefits of its success. Joining the ecosystem of an early-stage Charismatic Leader to develop foundational products and services for the smart product platform is an attractive option. If the ecosystem leader succeeds, early partners will be in an excellent position to define and deliver the next generation of smart products and services within the industry.

These considerations also apply to the Transformer ecosystem model. The main difference is that a well-established brand and large customer base are not as essential for a Transformer ecosystem to succeed. Transformers may base their strategy on an open smart product platform and focus on convincing as many partners as possible to develop products and services for that platform. Unlike the Charismatic Leader model, the Transformer ecosystem can be a good fit for small companies with advanced technology solutions. Chances of small company success are enhanced if it can quickly attract strong partners who share its goals of industry transformation. In many cases the transformational technology innovator becomes an acquisition target for these large companies. Success as a Transformer requires careful analysis of the strength of the incumbent leaders in a target industry sector, especially sectors traditionally dominated by Hegemon models. Given the state of Hegemon control of the energy sector, for example, would-be Transformers are unlikely to succeed in causing dramatic transformation in the short run. However, they may attract a number of like-minded partners to join a Transformer ecosystem that is targeting specific sectors in the early stages of disruption, if the Transformer can establish alliances that provide alternative paths to entry to previously closed markets. To accomplish this, the Transformer must be adept at making strategic alliances and at developing innovative business and pricing models.

Unless a company is rich in resources and prepared to nurture the Federator ecosystem during a very slow process of growth and adoption of the core technology platform, it will be almost impossible to succeed as the founder of a Federator model. However, as with the Charismatic Leader ecosystem, companies may benefit from developing products and services for an existing Federator platform, particularly if the Federator technology and partnerships facilitate entry to a new industry sector. There is little risk for smaller companies and start-ups wishing to leverage the technology provided by a successful Federator. And since the Federator is not attempting to disrupt industry-leader business models, it has considerable appeal for large companies. This makes the Federator strategy a viable evolutionary option for Hegemon companies seeing new markets through licensing their technology or providing infrastructure services in markets that are not their primary focus. Many large wireless carriers, for example, are positioning themselves as Federators by enabling enterprise M2M services.

The Hegemon ecosystem model is not a viable starting point for new entrants developing smart product strategies but it still offers options for companies that are already market leaders. Hegemons must choose between maintaining the industry status quo and forgoing the opportunities of market disruption and deliberately disrupting their competition with innovative smart services which will also cannibalize their own revenues. While many Hegemons (and their shareholders) will balk at the prospect of cannibalization, there will be others who decide to bet on the future value of smart services, just as some traditional market leaders transformed their business models to embrace the Internet and e-commerce even at the cost of short-term revenue losses.

Hegemons may also decide to take a different role in a new industry sector, either as a Charismatic Leader who maintains control of the ecosystem and takes market share from industry incumbents or as a Federator who leverages its technology platform to attract new partners and enable new services. It may also be attractive for smaller companies to partner selectively with existing Hegemon ecosystems, especially in highly regulated sectors such as medical services and energy generation, where options for short-term disruption are limited and alliances with entrenched Hegemon leaders are necessary for reaching customers. Where regulations and public safety risks are not an issue, however, a Federator partnership provides new entrants with more flexibility to develop innovative smart service strategies.

What else can managers do to turn smart product platforms and services into sustainable advantages for their companies? The creation and delivery of visible customer value should be the top priority. The potential for connected smart products to dominate their markets and sustain profitable business models is inextricably linked to the tangible value that these products deliver to consumers. Few managers would disagree with this proposition. However, most companies that have launched smart products in the past ten years have put other factors ahead of delivering clear consumer value. The next section discusses creating visible customer value as a competitive strategy for smart products and services.

## CREATING VISIBLE CUSTOMER VALUE

The discussion of smart product successes and failures in previous chapters highlights a set of smart product characteristics associated with consumer perceptions of value. Consumers perceive more value in products that put them in control of decisions about features and functionality and that offer flexibility in using the product in ways that are adapted to their personal context and lifestyle. At the same time, smart products must provide ease of use and reliable performance when delivering services and content to the user. There is a high value for both the customer and the vendor in delegated control, when consumers explicitly agree to rely on the smart product to optimize smart services on their behalf. Even when explicit permission is granted to the vendor to take control of certain product functions, it is important to provide regular visible feedback on the benefits that autonomous product operations are generating for the customer, whether this value is in terms of improved battery life of an electric vehicle, more personalized discount offers from local merchants, or smarter routing around traffic congestion.

These characteristics may seem straightforward and even obvious. Few smart product vendors would set out to design products that masked valuable features or that made it unlikely that users would willingly delegate control or share information with them. In practice, however, many smart products are ineffective in creating visible value for their owners and are designed to keep many aspects of the product and platform functionality used by the vendor and its smart ecosystem partners hidden from the user. To the extent that the product monitors customer behavior and collects and analyzes information about usage, that information-gathering activity and the insights that it provides to the vendor are typically invisible to the owner. Such smart products exhibit a strategy of information hoarding rather than transparency. As illustrated in Figure 10.2, an information-hoarding strategy limits the product owner's visibility to a small slice of the full product platform functionality.

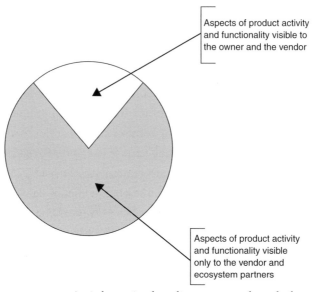

Aspects of product activity and functionality visible to the owner and the vendor

Aspects of product activity and functionality visible only to the vendor and ecosystem partners

FIGURE 10.2: An information-hoarding smart product platform

In contrast, the transparent smart product platform illustrated in Figure 10.3 makes the majority of the product's data gathering and ecosystem functionality visible to the customer in some way, perhaps in response to a proactive customer request to view such information on the smart product user interface or on the Web. The relative size of the visible and hidden sectors in each figure is not based on quantitative measures of specific product functionality; they are general indicators of visible and invisible activities based on the smart product data-collection capabilities and developer APIs discussed in earlier chapters.

Vendors may believe that they are acting in the customer's best interest when the majority of a smart product's functionality and data-collection activities work invisibly in the background. They would argue that consumers are often overwhelmed by too much information and that an overload of product functionality data would add to the perceived complexity of usage for most consumers. They might also believe that disclosing details about user monitoring and data collection would alienate some customers.

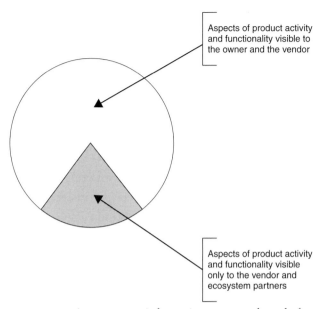

FIGURE 10.3: A transparent information smart product platform

The counter-argument is that designing smart product platforms and user interfaces to support user-controlled information
sharing fosters consumer trust and demonstrates the vendor's commitment to consumer privacy and transparency. The more that the
consumer trusts his product to act on his behalf, the more value the
product and related smart services can generate for both the owner
and the vendor. To the extent that most of a product's data-collecting
activities are hidden from the customer, claims about transparency
and disclosure are misleading at best. They may also backfire as customers who discover undisclosed smart product tracking activity are
likely to communicate that information to the media and to other
consumers via the Internet.

One way to establish transparency is to engage product owners
in an active feedback loop, motivating them to use the product's
embedded connectivity to communicate questions, suggestions,
and complaints directly to the vendor. Actively engaged owners
who want to explore all aspects of product functionality and are

interested in pushing the product to its limits may generate insights about the performance and potential value of the product that will be novel even to the product vendor. Such customers will often be the most willing to voluntarily share their experiences and opinions with the vendor and with fellow product owners. Vendors who decide to monitor their customers' behavior and collect data as a default option without explicitly surfacing this data collection to the product owner risk losing their customers' trust by trespassing on privacy expectations. In addition, vendors who rely on hidden monitoring forgo the opportunity to obtain deeper insights and increase customer loyalty through voluntary and transparent data collection that facilitates a dialog with product owners while respecting their privacy. Remote monitoring may play an important role in increasing enterprise operational efficiency and improving product support, but the product owner cannot appreciate that value unless it impacts their interaction with the product in some visible manner.

Even if many customers are not interested in the details of the product's embedded features, keeping them hidden implies that the vendor is not willing to be transparent. Making more of these activities visible to the customer is a strong statement that customers are important and respected members of the smart product ecosystem.

Another barrier to surfacing more information about product activities is the cost and effort that effective visualization and presentation of such information will require from the product designer and vendor. Conveying the details of product functionality in terms that a consumer can readily understand will be challenging, and will require innovative thinking about visualization of complex information. It will also require the vendor to examine a smart product's behind-the-scenes functionality, user monitoring, and data-collection activities more critically to determine if those features do in fact add value to either the product owner or the vendor. If the answer to both questions is "no" then eliminating such data collection would be a positive outcome for all.

The evolution of customer self-service provides an example of how motivated vendors who see clear economic and strategic benefits can design entirely new processes for customer interaction – and how customers can come to see increased value in these processes. Consumers are now accustomed to using ATM machines for banking transactions, to pumping their own gasoline at the self-service stations, and printing out boarding passes for airline flights at airport self-service kiosks or on personal printers. In the supermarket and big box stores, customers' attitudes have shifted from avoiding the self-service checkout lines to embracing them as the fastest and most efficient way to get from shopping cart to parking lot. More recently, airlines are providing options for us to download our boarding pass on our smartphones and ATM machines are making digital images of our deposit checks on the spot, eliminating even more traditional paper processing steps.

It has taken time for these models of self-service to become generally accepted and even preferred by customers. It has also taken years of enterprise investment in designing and deploying self-service interfaces that are intuitive enough to be accessible to the majority of consumers, and in many cases a further investment in educating consumers to use the self-service options efficiently, during which time company service representatives were on hand to provide assistance and answer questions. Creating a streamlined, effective process for customer self-service also required that companies carry out a careful review of their internal workflow. Before self-service interfaces could be offered to customers, companies often needed to streamline their internal processes and overall workflow, implementing technology to manage self-service customer data and transactions in real time.

Service providers spent time and money on self-service implementation because they believed that the long-term financial and competitive benefits of self-service would create an enormous return on their investment. Turning the work over to the customers allows a company to process more transactions with far fewer staff and at a higher level of customer satisfaction. It lowers overhead costs and makes internal infrastructure significantly more efficient and productive. These internal benefits would be compelling for many companies even if the self-service option had the effect of irritating or alienating some customers – much as the ever-present interactive voice response (IVR) systems for customer service on the phone do. But as it turns out, the majority of customers have come to prefer the self-service option if it is designed and implemented to give them a sense of control and to create genuine savings of time and effort for them as well as for the service provider. In a well-designed self-service system, customers see value in their interactions because they are in control of both the process and the outcome.

This combination of perceived efficiency and control is the key value proposition that makes a shift to self-service seem like a benefit rather than an added burden to customers. If the line at the self-serve pump or checkout were to be longer than the one at the staff-attended counter, the process would break down. But if we are in control of a process, and we see the self-service option as a free choice, we have a tendency to perceive it as being more efficient and faster than waiting for traditional service. Successful implementations of self-service operations take full advantage of this perception and create reinforcement to the customer sense of being in control by offering choices at appropriate points in the interaction. Some of these choices are highly profitable for the vendor as well as being of value to the customer, for example an offer to upgrade an airline seat at the point of printing out the boarding pass.

It has taken a decade of experience for this shift of responsibility for end-point services to take hold in many industries and companies have learned a great deal about consumer interaction in the process. The successful features of a customer self-service model are potential foundations for creating visible value in consumer interactions with their smart products, particularly when the owner has given the vendor permission to provide personalized offers and options based on past behavior.

Creating transparent smart product interfaces to give product owners more visibility into currently hidden activities of their smart products is not inherently more difficult or expensive than enabling customer self-service systems. If smart product ecosystem leaders are committed to information transparency, they will find ways to improve smart product design to deliver more visible value to their customers, including more options for unveiling details about product activities that are currently hidden from view.

Visible product value creates a dual bonus for the vendor. It fosters customer trust and it encourages customer engagement and data sharing. Consumer trust increases when a smart product performs as expected, conceals little, and allows owners to control when and how usage data is shared. As the popularity of social networks and user-generated content on the Web demonstrates, there are hundreds of millions of consumers who are comfortable with sharing information – as long as they get to control what data they share and can see personal value in the aggregation of that data. Providing visibility into more of a smart product's functions encourages interaction and information sharing among the groups of customers who are the most likely to become long-term product users and buyers of value-added services.

Smart products designed to encourage voluntary data sharing among customers will create strong customer loyalty as well as more information for the vendor about customer preferences. Encouraging the customer to create new value for the product through information disclosure to other product owners as well as to the vendor increases the visible value for all product owners. Smart services that depend on customer interaction and data sharing have the added competitive advantage of being very difficult to copy since they are based on customer trust and willingness to share personal data. No matter how advanced a smart product is on the day of its market launch, it is inevitable that competitors will match the product's functionality eventually, possibly at a lower price point. Since the value of smart services to the customer actually increases over time as the customer perceives that the service is becoming "smarter" based on repeated interactions, transparent smart products and services can create high barriers to customer exit. After a prolonged period of regular product use, it will be inconvenient or impossible for the owner to get a comparable level of service from a new product without investing time and effort in "breaking it in" to the same level of intelligence and personalization and, more importantly, building the same level of trust.

## CONCLUSION

We have seen that the macro trends in technology provide the impetus for embedding intelligence into a wider and wider spectrum of consumer products at all price points and connecting these products to each other and to the world at large. The increasing capacity and decreasing price of microprocessor chips and sensors, the spread of the Internet and ubiquitous connectivity via wireless networks, and the links between the billions of devices that comprise the Internet of Things are trends that will continue for the foreseeable future. Today we are in the early stages of an exciting cycle of smart product adoption, innovation, disruption, and transformation that will impact consumer lives and enterprise fortunes in the coming decade.

Up to this point, the technology of embedded intelligence and product connectivity has been far ahead of enterprise strategies for creating consumer value through smart services, or for leveraging smart product platforms to create profitable business models and sustainable competitive advantage. Many companies that invest in embedded product intelligence for consumer products expect that advanced technology will itself be enough of a value proposition to win consumer loyalty and market share. They are wrong. Consumers don't value embedded intelligence or product technology for its own sake; what matters is how well the product performs, what smart services it enables, and what visible new benefits it delivers.

By 2020 almost all products will be smart products in the sense of incorporating embedded intelligence and connectivity into their design. Even low-priced, commodity products will fit the definition we have used in this book, that smart products are network-connected consumer items with embedded microprocessors and software programmed to manage various aspects of the product's functionality. Since it will be taken for granted that embedded intelligence informs product behavior, the term "smart product" will no longer indicate any particular product differentiation. Managers will not expect this level of technology to be the basis for competitive advantage or market leadership. Some product companies will follow an enterprise services strategy and use embedded intelligence primarily to create internal operational efficiencies by reducing the costs of product maintenance and support. Other companies will try to use the hidden data-collecting and monitoring capabilities of smart products to extend their control over consumer behavior and product usage after sale.

Companies that aspire to market leadership will create smart product platforms that transform industry sectors with smarter services that deliver visible value to consumers. They will create profitable and scalable businesses for themselves and their ecosystem partners. These companies will be the leaders of tomorrow's smart product ecosystems.

# References

Abbott, Ralph E. 2005. "Time-of-Use Rates: Sideburns and Bellbottoms?" *Energy Markets*, July/August.

Abelson, Jenn. 2009. "A Once-great Affair Stuck in Park," *Boston Globe*, July 12. p. A11.

ABI Research. 2008a. ABI Research Report, November 13.

ABI Research. 2008b. Press Release, April 2.

Admob Report. 2009. Report available at www.admob.com

AHS. 1997. "Demo '97: Proving AHS Works," July/August, Vol. 61. No. 1, available at www.tfhrc.gov/pubrds/July97/demo97.htm

Allmendinger, Glen and Ralph Lombreglia. 2005. "Smart Strategies for the Age of Smart Services," *Harvard Business Review*, October, Vol. 83, No. 10.

Amazon. 2009. Amazon Kindle: License Agreement and Terms of Use, February 9, available at www.amazon.com/gp/help/customer/display.html?ie=UTF8&nodeId=200144530

American Beverage Institute. 2009. "Highway Bill Eliminates States' Authority on Interlock Use," June 19, available at www.abionline.org/pressRelease_detail.cfm?release=286.

America's Carriers Telecommunication Association. 1996. "In the Matter of the Provision of Interstate and International Interexchange Telecommunications Service via the 'Internet' by Non-Tariffed, Uncertified Entities," Petition before the FCC, March 4.

Anderson, Nate. 2009. "'Anonymized' Data Really Isn't – and Here's Why Not," *Ars Technica*, September 8, available at http://arstechnica.com/tech-policy/news/2009/09/your-secrets-live-online-in-databases-of-ruin.ars

Ankeny, Jason. 2009a. "Apple License Update Bans Rogue iPhone Apps," *FierceMobileContent*, April 3, available at www.fiercemobilecontent.com/story/apple-license-update-bans-rogue-iphone-apps/2009-04-03

Ankeny, Jason. 2009b. "Counting down the iPhone's hottest paid apps," *FierceWireless*, October 15, available at www.fiercemobilecontent.com/story/counting-down-iphones-hottest-paid-apps/2009-10-15

Arino, Mitsuo. 2007. "ITS Policy in Japan and SmartWay," ITS Policy and Program Office. Government of Japan, October.

Asia Finest Discussion Forum. 2009. "Younger Japanese Not Interested in Cars," February 21, cited from www.asiafinest.com/forum

Askland, Andrew. 2007. "The Double Edged Sword that is the Event Data Recorder," *Temple Journal of Science*, Technology & Environmental Law, Vol. XXV, No. 1.

Austin Energy. 2009. "Austin Energy Updates Solar Rebate Program," October 1, available at www.austinenergy.com/About%20Us/Newsroom/Press%20Releases/2009/solarRebate. htm

Baker, William B. 2009. "Does the FTC Action Against Sears Cast Doubt on the Benefit of Website Privacy Policies?" *Privacy in Focus*, October, available at www.wileyrein.com/publications.cfm?sp=articles&id=5595#

Barron's. 2009. "Amazon: $2B/Yr Kindle Sales by 2012," *Barron's*, June 4, available at http://blogs.barrons.com to subscribers.

Belknap, Kevin. 2009. Author interview, August 20.

Benedict, David. 2007. "Human Factor Issues of Driving Assistance Systems," Forum on Advanced Vehicle Safety Technology, January.

Berst, Jesse. 2009. "Why Control4 Is the Company to Beat in Home Energy Management," *SmartGrid News*, July 6.

Bethell, Catherine. 2009. "Mobile Open Source Operating Systems: Markets & Opportunities 2009–2014," Juniper Research, July.

Better Place. 2009. Available at www.betterplace.com

Black, Sam. 2002. "Is Toro's iMow Cutting it?" *Minneaopolis St. Paul Business Journal*, April 5, available at www.bizjournals.com/twincities/stories/2002/04/08/story2.html.

Blincoe, L., A. Seay, E. Zaloshnja, T. Miller, E. Romano, S. Luchter, and R. Spicer. 2002. "The Economic Impact of Motor Vehicle Crashes 2000," National Highway Traffic Safety Administration, DOT HS 809 446, May.

BMW. 2008. "The New 5th Generation BMW 7 Series Breaks Cover," BMW US Press Information, July 5, available at https://resource.bmwusa.com/pdf_64efcb5d-9645–4152-a794-dffeaf8dd46c.arox

Bohn, Dieter. 2009. "FYI: The Pre Reports Your Location to Palm," August 12, available at www.precentral.net/fyi-pre-reports-your-location-palm

Bordoff, Jason E. and Pascal J. Noel. 2008. "Pay-As-You-Drive Auto Insurance: A Simple Way to Reduce Driving-Related Harms and Increase Equity," Brookings Institute, July.

Brisbourne, Alex. 2009. Author interview, July.

Brockway, Nancy. 2008. "Advanced Metering Infrastructure: What Regulators Need to Know About Its Value to Residential Customers," National Regulatory Research Institute, February 13.

Business Wire. 2004. "P2P MMS Is Expected to Replace P2P SMS as the Leading Messaging Revenue Generator by 2008," November 19, available at http://findarticles.com/p/articles/mi_m0EIN/is_2004_Nov_19/ai_n6367662/

Business Wire. 2009. "TomTom Unveils Connected Portable Navigation Device in the U.S.," January, available at www.businesswire.com/news/google/20090108005548/en

CABA. 2008. "2008 State of the Connected Home Market Study," available to download at www.caba.org

CABA. 2009. "Research Shows that the Possibilities are Endless, but Convergence is Key," available at www.caba.org

CardioNet. 2009. "CardioNet Announces Highmark Medicare Services Reimbursement Reduction Regarding CPT Code 93229," CardioNet Press Release, July 12.

Chambers, Doug and Svetlana Grant. 2009. "AT&T: Preparing for a Tidal Wave of Emerging Devices," GSMA Embedded Mobile Newsletter, September, available at www.gsmworld.com/documents/embedded_mobile/emedded_mobile_sep09_full.pdf

Christensen, Clayton M. 1997. *The Innovator's Dilemma*, Watertown, MA: Harvard Business Press.

CNET News. 1998. "Lucent Leans on Home Networking," October 6, available at http://news.cnet.com

Connolly, Christine. 2009. "Driver Assistance Systems aim to Halve Traffic Accidents," *Sensor Review*, Vol. 29, No. 1.

Continua. 2009. Newsletter, August 27.

Control4. 2009. August, available at www.control4.com

Cory, Karlynn, Toby Couture, and Claire Kreycik. 2009. "Feed-in Tariff Policy: Design, Implementation, and RPS Policy Interactions," Technical Report NREL/TP-6A2-45549, March, available at www.nrel.gov/docs/fy09osti/45549.pdf

Cronin, Mary J. 2000. "Privacy and Electronic Commerce," *Public Policy and the Internet: Privacy, Taxes, and Contract*, edited by Nicholas Imparato, Stanford: Hoover Press.

CSA International. 2009. May 29, available at www.csa-international.org

CTIA. 2009. CTIA Semi-Annual Wireless Industry Survey, June.

DARPA. 2009a. DARPA Urban Challenge, available at www.darpa.mil/grandchallenge/index.asp

DARPA. 2009b. DARPA Urban Challenge Overview, available at www.darpa.mil/grand-challenge/overview.asp

Databeans. 2009. "2009 Market Research Report," available at www.databeans.net/reports/current_reports/SemiProd_Microcontrollers.php.

DataMonitor. 2009. "GE and Intel Invest $250 Million in New Market Opportunity," eHealth News.eu, May 12, available at www.ehealthnews.eu

Davis, Doug. 2009. "Welcome to the Embedded Internet, on the Threshold of a New World," July 19, available at www.intel.com/embedded/15billion/embedded-internet/threshold-new-world.htm.

Davis, Jim and Michael Kanellos. 1999. "Microsoft Touts Home Networking Standard," *CNET News*, January 8, available at http://news.cnet.com

Design News. 2008. "TI Rolls Out Low-Power Wireless Microcontroller Platform," December.

Dickson, Frank. 2009. "Mobile VoIP – Transforming the Future of Wireless Voice," *In-Stat*, September.

Duncan, Roger. 2009. Author interview, February 20.

Echelon Corporation. 2009. Poly-Phase Brochure, available at www.echelon.com/products/allproducts.htm

Electronic Frontier Foundation. 2010. "Unintended Consequences: Twelve Years Under the DMCA," available at www.eff.org/wp/unintended-consequences-under-dmca

Electronic Privacy Information Center. 2009. "Re-identification," available at www.epic.org

Electro to Auto Forum. 2009. "Electronics to Account for 40% of Automotive Production Costs by 2015," available at http://e2af.com/trend/071210.shtml

Ember. 2009. Available at www.ember.com/zigbee_ember_zigbee.html

Energy Information Administration. 2008. "The International Energy Outlook." Available at www.eìa.doe.gov

Environmental Protection Agency. 2009a. "On-Board Diagnostics (OBD): Basic Information," available at www.epa.gov/otaq/regs/im/obd/basic.htm

Environmental Protection Agency. 2009b. "On-Board Diagnostics (OBD): Frequent Questuion," available at www.epa.gov/otaq/regs/im/obd/questions.htm#9

Evans, Christopher. 1979. *The Micro Millennium*, New York, NY: Viking Press, book jacket copy.

FARS Encyclopedia. 2009. Available at www.fars.nhtsa.dot.gov/main/index.aspx

Federal Energy Regulatory Commission. 2006. "Assessment of Demand Response and Advanced Metering," Staff Report, Washington, DC, August.

Federal Energy Regulatory Commission. 2009. "A National Assessment of Demand Response Potential," Staff Report, Washington, DC, June.

Federal Trade Commission. 2009. "In the Matter of Sears Holdings Management Corporation," Docket No. C-4264a, August 31.

Foley, Mark. 2008. "The Dangers of Meter Data (Part 1)," *Smart Grid News*, June 2, available at www.smartgridnews.com

Fox-Penner, Peter S., Marc W. Chupka, and Robert L. Earle. 2008. "Transforming America's Power Industry: The Investment Challenge," The Brattle Group, April 21.

Gantz, John. 2009. "The Embedded Internet: Methodology and Findings," *IDC*, January, cited by Doug Davis on the Intel website.

Garmin. 2009. Available at www.garmin.com

General Electric. 2009. "GE Targets Net Zero Energy Homes by 2015," Press Release, July 14, available at www.genewscenter.com/content/detail.aspx?releaseid=7272&newsareaid=2.

GM Media OnLine. 2009. Cited from www.media.gm.com/us/onstar/en/company/history/history.html.

Google. 2009. "Google PowerMeter Privacy Policy Notice," Google, available at www.google.com/powermeter/privacy

Green, Heather. 2009. "The Static Over Smart Grids," *Business Week Online Energy*, April 2, available at www.businessweek.com/magazine/content/09_15/b4126048296127.htm

Gyorke, Shawn. 2008. "Deciphering the NHTSA Event Data Recorder Ruling," available at www.crashdataservices.net

Guthery, Scott and Mary Cronin. 2003. *MMS Application Development*, New York: Macmillan.

Halderman, J. Alex and Edward W. Felten. 2006. "Lessons from the Sony CD DRM Episode," February 14, available at www.copyright.gov/1201/2006/hearings/sonydrm-ext.pdf

Hall, Ben. 2009. "Hachette Chief Laments Digital Pricing Threat to Book Publishing," *Financial Times*, August 31, available at www.ft.com to registered users and subscribers.

Hansell, Saul. 2009. "Verizon Boss Hangs Up on Landline Phone Business," September 17, *New York Times*, Bits, available at http://bits.blogs.nytimes.com/2009/09/17/verizon-boss-hangs-up-on-landline-phone-business/?apage=3

HealthVault. 2009. Available at www.healthvault.com.

Hebner, Scott. 2009. "Smarter Products," IBM, May.

Heskett, Ben. 1999. "1999: The Year for Home Networking?" *CNET News*, available at http://news.cnet.com

Honan, Mathew. 2009. *Wired*, January 19.

Honda. 2009. "Honda to Exhibit its Latest Advanced Safety Vehicles," Honda Motor Co., Ltd, Press Release, February 19, available at http://world.honda.com/news/2009/4090219ITS-Safety-2010

Hug, Ralf. 2009. Author interview, July.

IAB. 2009. "Economic Value of the Advertising-Supported Internet Ecosystem," August, available at www.iab.net/insights_research/530422/economicvalue

Inouye, Susie. 2008. "Car Safety Systems and the Chips that Serve," *EE Times Asia*, January 16, available at www.eetasia.com for registered users.

Intel. 2009a. "Intel's First Microprocessor – the Intel® 4004" Intel Corporate Archives, available at www.intel.com/museum/archives/index.htm.

Intel. 2009b. "Moore's Law", available at www.intel.com/technology/mooreslaw/.

Intel. 2009c. "20 Years – Intel: Architect of the Microcomputer Revolution," available at www.intel.com/museum/archives/brochures/brochures.htm

iSuppli. 2009a. Magney, P. "BMW 7 Series Tops iSuppli's Automotive Technology Scorecard," *iSuppli News*, April 2, available at www.isuppli.com

iSuppli. 2009b. Magney, P. "OnStar Shines for GM, as Auto Telematics Becomes Mainstream," *iSuppli News*, April 28, available at www.isuppli.com

ITS Strategy Committee. 2003. "ITS Strategy in Japan, Report of the ITS Strategy Committee," Summary version, July.

ITS World. 2001. "ITS the SmartWay," March – April, Vol. 6, No. 2.

J. Arnold & Associates. 2009. *Telecommunications Industry Issues Index.*

Jacobson, Julie. 2009. "All Whirlpool Appliances to be 'Smart' by 2015," *Electronic House*, May 6, available at www.electronichouse.com/article/ all_whirlpool_appliances_to_be_smart_by_2015/C157.

Jajah. 2008. "Gizmo5 Joins Yahoo! and MailVision in Choosing JAJAH's Managed Services Platform to Drive Global Growth," Press Release, June 26, available at www.jajah.com

Kanellos, Michael. 2004. "South Korea's House of the Future," *CNET News*, May 13, available at http://news.cnet.com

Kendrick, James. 2009. "Verizon: Please Stop Disabling GPS in Smartphones on Your Network," *JK on the Run*, July 17, available at http://jkontherun.com/2009/07/17/verizon-please-stop-disabling-gps-in-smartphones-on-your-network/

Kramer, Staci. 2009. "Microsoft Narrows Device Options with Zune HD," September 15, available at http://paidcontent.org

Krazit, Tom. 2009. "Google's Rubin: Android 'a Revolution'," May 21, *CNET News*, available at http://news.cnet.com/8301-1023_3-10245994-93.html

Kurzweil, Ray. 2005. *The Singularity is Near*, New York, NY: Viking Press, available at http://singularity.com/bookexcerpts.html

Kwun, Michael. 2008. "EFF Opposes MPAA's Selectable Output Control FCC Petition," *Electronic Frontier Foundation*, July 22, available at www.eff.org/deeplinks/archive

LaReau, Jamie. 2009. "GM offers OnStar to other Automakers," *Automotive News*, February 2, available at www.autonews.com

Largent, Steve. 2009. "CTIA Statement on FCC Open Commission Meeting," *CTIA*, August 27, available at www.ctia.org/blog/index.cfm/2009/8/27/CTIA-Statement-on-FCC-Open-Commission-Meeting

Lasar, Matthew. 2008. "Any Lawful Device: 40 Years After the Carterfone Decision," *Ars Technica*, June 26, available at www.arstechnica.com/tech-policy/news/2008/06/carter-fone-40-years.ars

Lessig, Lawrence. 1999. *Code and Other Laws of Cyberspace*, New York, NY: Basic Books.

Lessig, Lawrence. 2009. "Continuing the Work of Code," *Cato Unbound*, May, available at www.cato-unbound.org/2009/05/11/lawrence-lessig/Continuing-the-Work-of-code.

Leveson, Nancy and Clark S. Turner. 1993. "An Investigation of the Therac-25 Accidents," *IEEE Computer*, Vol. 26, No. 7.

LG Electronics. 2002. "LG Internet Refrigerator is at The Heart of The Digital Home Network," Press Release, available at www.prnewswire.com/cgi-bin/stories.pl?ACCT= 104&STORY=/www/story/01-13-2002/0001646820&EDATE=)

LG Electronics. 2009. "Home Network Technologies," available at www.lg.com/us/about-lg/ innovation/technology/home-network-technologies.jsp

Lockton, Daniel. 2005. "Architectures of Control in Consumer Product Design," M.Phil. Technology Policy, June, Judge Institute of Management, University of Cambridge.

Logitech. 2009. Available at www.logitech.com/index.cfm/remotes/universal_remotes/devices/4708&cl=us,en

Lyardet, F. and E. Aitenbichler. 2008. *Smart Products: Building Blocks of Ambient Intelligence*, New York, NY: Springer.

M2M Magazine. 2008. "Is It Touching Our Everyday Lives?" Nov/Dec, available at www.m2mmag.com/issue_archives/default.aspx?ID=697

M2M Premier. 2009. M2M Premier Glossary, available at www.m2mpremier.com/Technology_glossary.aspx?id1=S to subscribers.

MacDailyNews. "Do not unlock iPhones; many unlocking programs cause irreparable iPhone software damage," MacDailyNews, September 24, available at www.macdailynews.com/index.php/weblog/comments/14991/

Malik, Om. 2009. "Some Fun Facts About The Google Phone," GigaOm, www.gigaom.com/2009/04/02/how-t-mobile-customers-use-the-google-phone

Martin, Nick. 2009. "What's Next for Electronics Devices?" *Altium*, April, available at the Resource Center section of www.altium.com.

Mazda. 2009. "Mazda at the 41st Tokyo Motor Show," Mazda, Press Release, October, available at www.Mazda.com/publicity/release/2009/200902/090217a.html

McGrath, Michael E. 1995. *Product Strategy for High-Technology Companies*, Burr Ridge, IL: Irwin.

McMillan, Alex. 2002. "DoCoMo's i-mode Heading to America," *CNN*, May, available at http://edition.cnn.com (using the search function).

Meeker, Mary. 2009. "Economy + Internet Trends," *Morgan Stanley presentation* at the Web2.0 Summit, October 20, and Berg Insight Report, BER40, April.

Microsoft. 2007. "New Study Finds 14.7 Million Jobs Created Globally by Microsoft and Its Ecosystem."

Microsoft. 2009a. Microsoft HealthVault Ecosystem, available at www.healthvault.com/Industry/index.html.

Microsoft. 2009b. Microsoft Service Agreement Supplement Updated, June, available at www.microsoft-hohm.com to subscribers.

Migliore, Greg. 2009. "OnStar is Not for Sale," AutoWeek, September 9, available at www.autoweek.com/article/20090908/CARNEWS/909089991

Miles, Stephanie. 1999. "Cisco Details its Home Network Plans," *CNET News*, January 7, available at http://news.cnet.com

Miller, Robert. 2009. "Is Deutsche Telekom Playing an April's Fool Joke at the Expense of Skype Users in Germany?" *The Big Blog*. April 1, available at http://share.skype.com/sites/en/2009/04/is_deutsche_telekom_playing_an.html

Milne, George R. and Mary J. Culnan. 2004. "Strategies for Reducing Online Privacy Risks: Why Consumers Read (or Don't Read) Online Privacy Notices," *Journal of Interactive Marketing*, Vol. 18, No. 3.

Minarini, Fabrizio. 2002. "The European Framework for the Development of Intelligent Vehicles," The eSafety Initiative, available at www.awake-eu.org/pdf/ec_framework. pdf

Mobile Marketing Association. 2008. Global Code of Conduct, July 15.

Morrison, Dianne See. 2009a. "German Carrier T-Mobile Blocking Skype," April 1, available at www.moconews.net/entry/419-german-carrier-t-mobile-blocking-skype

Morrison, Dianne See. 2009b. "T-Mobile Germany: Pay for VoIP Usage Or We Will Continue to Block it," June 4, available at www.moconews.net/entry/419-t-mobile-germany-pay-for-voip-usage-or-we-will-continue-to-block-you

Mothers Against Drunk Driving. 2009. Available at www.madd.org/Drunk-Driving/Drunk-Driving/Statistics.aspx

Mulligan, Deirdre K., Jack I. Lerner, Erin Jones, Jen King, Caitlin Sislin, Bethelwel Wilson, and Joseph Hall. 2009. "Samuelson Law Privacy and the Law in Demand Response Energy Systems," Technology & Public Policy Clinic, University of California, Berkeley, cited from www.samuelsonclinic.org

Murray, Charles J. 2005. *EE Times*, June 5.

National Institute of Standards and Technology. 2009. "NIST Framework and Roadmap for Smart Grid Interoperability Standards, Release 1.0," September, available at www.nist.gov/public_affairs/releases/smartgrid_interoperability.pdf

Natsuno, Takeshi. 2003. *The i-mode Wireless Ecosystem*, London: John Wiley. Originally published in 2002 by Tokyo: Nikkei BP Planning, Inc.

Nespresso. 2009. Available at www2.nespresso.com/mediacenter/index.php?page=about-us

Newcomb, Doug. 2009. "Automaker vs. Aftermarket Tech: Which Offers Better Value?" May 7, available at www.edmunds.com/ownership/audio/articles/147452/article.html

New York Times. 2009. "An Interview with David Vladeck of the FTC" *New York Times*, August 5, available at http://mediadecoder.blogs.nytimes.com/2009/08/05/an-interview-with-david-vladeck-of-the-ftc/?scp=1&sq=vladeck%20ftc%20privacy&st=cse

Nissan. 2009. Cited in "Nissan Participates in ITS-Safety 2010 Industry-Wide Test," Nissan Motor Co. Ltd, Press Release, January 7, available at www.nissan-global.com/EN/NEWS/2009/_STORY/090107–01-e.html

Nokia. 2000. "Nokia and KPN Co-operate to Create Home of the Future," Press Release, February 10.

Nokia. 2009. Available at www.nokia.com/about-nokia/financials

NPD Group. 2009. "iTunes Leads with 25 Percent of all Music Sold," *News Release*, August 18, available at www.npd.com/press/releases/press_090818.html

NTT DoCoMo. 2009. i-mode History and Timeline, available at www.nttdocomo.com/services/imode/history/index.html

Nuttall, Chris. 2009. "Google Voice on Mobile is a Warning Call," July 15, *Financial Times* Technology blogs, available at http://blogs.ft.com/techblog/author/chrisnuttall

OBD-Codes.com. 2009. Cited from OBD-Codes.com

O'Brien, Danny. 2009. "License to Kill Innovation: the Broadcast Flag for UK Digital TV?" Electronic Frontier Foundation, September 14, available at www.eff.org/deeplinks/archive

OnStar. 2009. Available at www.onstar.com.

Oracle. 2009. "Turning Information into Power," March 9.

Ostrow, Adam. 2007. "The 7 Most Disruptive VoIP Services," July 3, available at http://mashable.com/2007/07/03/voip

Palm. 2009. Available at www.palm.com

Pew Research Center. 2009. Internet & American Life Project, available at www.pewinternet.org

Philips Research. 2009. HomeLab, available at www.research.philips.com/technologies/

Philips, T. 2009. Cited from "The Updated 2009 Mercedes-Benz S-Class Makes Its World Debut," April 9, available at www.emercedesbenz.com/Apr09/09_001633_The_Updated_2009_Mercedes_Benz_S_Class_Makes_Its_World_Debut.html

Plamondon, Scott. 1999. "Jini, Sun's Magic out of the Lamp," *JavaWorld.com*, February 1, available at www.javaworld.com

Proctor, Cathy. 2009. "Xcel Drops Proposed Solar-power Fee," *Denver Business Journal*, August 5, available at http://denver.bizjournals.com/denver/stories/2009/08/03/daily31.html

Progressive. 2009. MyRate FAQ, cited from www.progressive.com/myrate/myrate-faq.aspx

Pure Mobile. 2009. "The Origins of Camera Phones," available at www.puremobile.com/cameraphones.asp

Qualcomm. 2009. "What Are Smart Services?" August, available at www.qualcomm.com/products_services/mobile_content_services/enterprise/smartservices/smartservices.html

Radin, Margaret Jane. 2004. "Regulation by Contract, Regulation by Machine," *Journal of Institutional and Theoretical Economics*, Vol. 160, pp. 1–15.

Ramsay, Maisie. 2009. "Carriers Warming Up to Wi-Fi," *Wireless Week*, August 27.

RedLaser. 2009. Cited from www.redlaser.com/SDK.aspx

Rijsdijk, S. A., E. J. Hultink, and A. Diamantopoulos. 2007. "Product Intelligence: Its Conceptualization, Measurement and Impact on Consumer Satisfaction," ERIM Report Series Reference No. ERS-2007–006-ORG, available at http://ssrn.com/abstract=1303356.

Rose, David. 2009. Author interview, August 13.

Rosenthal, Morris. 2009. "2008 US Sales for Amazon, BN.com, Barnes & Noble and Dalton, Borders and Waldenbooks," cited from www.fonerbooks.com/booksale.htm

Safety Monitor. 2009. "Which Road Safety Targets to Guide the EU to 2020," Editorial, No. 78, October.

Salonen, Petri I. 2004. "Evaluation of a Product Platform Strategy for Analytical Application Software," Doctoral Thesis, Helsinki School of Economics.

Samsung. 1998. Samsung Press Release and *Wired* magazine, July 1.

Savitz, Eric. 2009. "Cowen Sees 3M Kindle Users by Year-end 2010," *Barron's*, August 10, available at http://blogs.barrons.com to subscribers.

Scheck, Justin. 2009. "Accusations of Snooping in Ink-Cartridge Dispute," *Wall Street Journal*, August 11.

Sisario, Ben. 2009. "Sales of CDs Continue Decline in 2008," *New York Times*, Technology Section, January 1, available at www.nytimes.com/2009/01/01/technology/01iht-music.4.19032321.html

Skype. 2009. Available at www.skype.com

Skype Communications. 2007. "Petition to Confirm a Consumer's Right to Use Internet Communications Software and Attach Devices to Wireless Networks, RM11351," February 20.

Skype Communications. 2008. "Comment in the Matter of Exemption to Prohibition on Circumvention of Copyright Protection Systems for Access Control Technologies, RM2008–08," August 28.

Smart Start. 2009. Cited from www.smartstartinc.com

Sokol. D. Daniel. 2001. "The European Mobile 3G UMTS Process: Lessons from the Spectrum Auctions and Beauty Contests," *Virginia Journal of Law & Technology*, Vol. 6.

Solheim, Shelley. 2005. "Verizon Wireless Users Sue Over Disabled Bluetooth Features," *eWeek.com*, January 14, available at www.eweek.com/c/a/Mobile-and-Wireless/Verizon-Wireless-Users-Sue-Over-Disabled-Bluetooth-Features/

Stanford. 2009. Available at http://design.stanford.edu/spdl/.

Steimel, George. 2009. Consumer M2M Report, Beecham Research, available at www.beechamresearch.com/report.aspx?reportid=23

Steinbuch, M. 2009. Available at www.elsevier.com/wps/find/journaldescription.cws_home/933/description#description

Sviokla, John. 2005. "In Praise of Ecosystems," *Fast Company*, available at www.fastcompany.com/magazine/97/open_essay.html

Swindle, Orson. 1999. "An FTC Commissioner Looks at Internet Privacy," Federal Trade Commission Privacy In American Business Conference, November, Arlington, VA, available at www.ftc.gov/speeches/swindle/westin.shtm

Symbian Foundation. 2009. Available at www.symbian.org

TDG. 2008. "Your Next Video Service May Come From Your Gaming Console," Press Release, December 1. www.tdgresearch.com/blogs/press-releases/archive/2008/11/30/tdg-report-your-next-video-service-may-come-from-your-gaming-console.aspx

THOMAS. 1998. "Section 1201 Circumvention of Copyright Protection Systems," Digital Millennium Copyright Act (Enrolled as Agreed to or Passed by Both House and Senate), H. R. 2281, available at http://thomas.loc.gov

TomTom. 2009. Available at www.tomtom.com.

US Department of Energy. 2006. "Benefits of Demand Response in Electricity Markets and Recommendations for Achieving Them," *Report to Congress*, February, available at www.oe.energy.gov/DocumentsandMedia/congress_1252d.pdf

US Department of Energy. 2007. "Appendix A5: A Systems View of the Modern Grid, Conducted by the National Energy Technology Laboratory for the Office of Electricity Delivery and Energy Reliability," January, available at www.netl.doe.gov/moderngrid/docs

US Department of Energy. 2008a. "Knowledge Brought to Power," Report No. 20.

US Department of Energy. 2008b. Cited from "The Modern Grid Strategy, Characteristics of the Modern Grid 2008," *National Energy Technology Laboratory* (NETL) Department of Energy, available at www.netl.doe.gov/moderngrid

US Department of Energy. 2009a. "The Smart Grid: An Introduction," *Department of Energy*, available at www.oe.energy.gov/1165.htm

US Department of Energy. 2009b. "The Effect of Income on Appliances in U.S. Households," Energy Information Administration, 2001 Residential Energy Consumption Survey, available at www.eia.doe.gov/emeu/recs/appliances/appliances.html

US Department of Transportation. 2008. "National Motor Vehicle Crash Causation Survey," *US Department of Transportation*, DOT HS 811 059, July.

US Department of Transportation. 2009a. Research and Innovative Technology Administration, available at www.its.dot.gov/faqs.htm

US Department of Transportation. 2009b. DOT Press Release, available at www.rita.dot.gov/publications/horizons/2009_05_01/html/connectivity_the_evolving_paradigm_for_intellidrive.html

US Federal Communications Commission. 2003. "FCC Clears Way for Local Number Portability Between Wireline and Wireless Carriers," November 10, available at http://fjallfoss.fcc.gov/edocs-public/attachmatch/DOC-241057A1.pdf

UtiliPoint. 2008. Press Release, July 31.

Valocchi, Michael, John Juliano, and Allan Schurr. 2008. "IBM Lighting the Way: Understanding the Smart Energy Consumer," Global Business Services.

van Gerwen, Rob, S. Jaarsma and R. Wilhite. 2006. "Smart Metering," Distributed Generation, KEMA, The Netherlands, available at www.leonardo-energy.org/webfm_send/435

Verizon Wireless. 2008. "Wireless Motorola V710 Settlement Claim Filing Instructions," available at www.verizonwireless.com/pdfs/v170settlement/V710_Claim_form_and_instructions.pdf

Vladeck, David C. 2009. "Promoting Consumer Privacy: Accountability and Transparency in the Modern World," October 2, available at www.ftc.gov/speeches/vladeck/0901002nyu.pdf

Vodafone. 2009. "Vodafone Mobilizes For Fast Growing M2M Market," Press Release, July 21.

Volvo. 2008. "Volvo S60 Concept: A Preview of the 2010 Volvo S60," Dec 16, cited from http://jalopnik.com

Volvo Technology Magazine. 2008. "Alcolock development gathers pace," September 15, available at www.volvo.com/group/global/en-gb/productsandservices (under Research & technology and News & updates).

Wagoner, Rick. 2008. "Electric Avenue – The Convergence of Electronic and Automotive Technologies," *Consumer Electronics Show* – Las Vegas, NV, January 8, available at http://archives.media.gm.com/archive/documents/domain_4/docId_42568_speech.html

Wayner, Peter. 2009. "iPhone App Store Roulette: A Tale of Rejection," *Computerworld Development*, available at www.computerworld.com/s/article/9135664/iPhone_App_Store_roulette_A_tale_of_rejection?taxonomyId=63&pageNumber=2

Welsh, Jonathan. 2009. "Late on a Car Loan? Meet the Disabler," March 25, *Wall Street Journal*, available at http://online.wsj.com/article/SB123794137545832713.html

Williams, Mike, Adriana Blanco, Joseph Unsworth, Masatsune Yamaji, Jon Erensen, and Stan Bruederle. 2008. "Hype Cycle for Automotive Electronics, 2008," Gartner Research, August.

Wired. 2003. O'Brien, Jeffrey M. "Bill Gates, Entertainment God," July, available at www.wired.com/wired/archive/11.07/40gates.html

Wireless Expertise. 2009. "The Future of Mobile Application Storefronts," *Wireless Expertise*, September, available at http://wirelessexpertise.com/research.php

Wood, Nick. 2009. "Worldwide Smartphone Sales to Outstrip Notebook Sales in 2009," *Total Telecom*, October 27.

Xcel Energy. 2009. Available at www.xcelenergy.com/minnesota/Company/AboutUs/Pages/Temp.aspx

Zaret, Elliot. 2003. "Access Denied: The Limits of Fair Use," *DC Bar*, February, available at www.dcbar.org/for_lawyers/resources/publications/washington_lawyer/february_2003/access.cfm

Zetter, Kim. 2009. "Feds at DefCon Alarmed After RFIDs Scanned," *Wired.com*, August 4, available at www.wired.com/threatlevel/2009/08/fed-rfid/

Zittrain. 2008. *The Future of the Internet – And How to Stop It*, New Haven, CT: Yale University Press.

# Index